The Cambridge Manuals of Science and
Literature

THE MIGRATION OF BIRDS

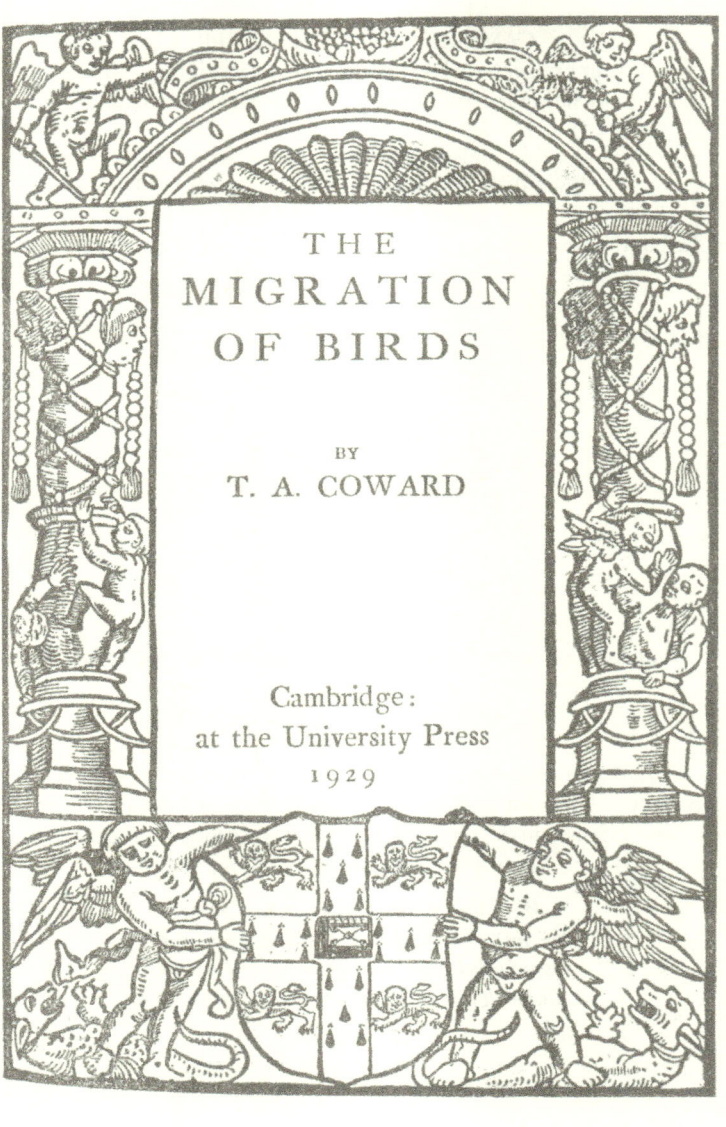

THE MIGRATION OF BIRDS

BY

T. A. COWARD

Cambridge:
at the University Press

1929

CAMBRIDGE UNIVERSITY PRESS
Cambridge, New York, Melbourne, Madrid, Cape Town,
Singapore, São Paulo, Delhi, Tokyo, Mexico City

Cambridge University Press
The Edinburgh Building, Cambridge CB2 8RU, UK

Published in the United States of America by Cambridge University Press, New Yo

www.cambridge.org
Information on this title: www.cambridge.org/9781107606098

First published 1929
First paperback edition 2011

First Edition, 1912
Second Edition, 1912
Third Edition, 1929

A catalogue record for this publication is available from the British library

ISBN 978-1-107-60609-8 Paperback

*With the exception of the coat of arms at
the foot, the design on the title page is a
reproduction of one used by the earliest known
Cambridge printer, John Siberch, 1521*

PREFACE

ANY attempt to elucidate the problems connected with the Migration of Birds must, in the present state of knowledge, contain some theory and speculation, but the diligent observations of an army of careful workers yearly add facts, which though they may appear insignificant when considered alone, tend in the aggregate to confirm or repudiate the conclusions of past workers. I have endeavoured to bring together some of the more important theories, and to give prominence to ascertained facts ; I have also striven to check desire on my own part to wander into realms of pure speculation, though conscious that I have not always evidence to support my suggestions.

The numbers in brackets ( ) in the text refer to the books or papers mentioned in the list at the end of the volume, which is in no ways an attempt at a full bibliography. I have quoted freely from the works of past and living ornithologists. To these I offer apologies if I have misconstrued their arguments, and acknowledge my indebtedness to those whose observations or writing

have given me light.   In particular I tender thanks
to Mr Wells W. Cooke for his permission to reproduce the maps facing pp. 76, 78, 80.   I have
found his writings and those of Herr Otto Herman
and Mr W. Eagle Clarke especially valuable.   Mr
Eagle Clarke's long looked-for book on Migration is,
as I write, still in the press ; had mine been more
than a manual I should have hesitated to publish
until his had appeared.

T. A. COWARD.

BOWDON, CHESHIRE,
4 *November* 1911.

PREFACE TO THIRD EDITION

THE advance in aviation teaching new facts about
the upper air, the observations of experienced airmen,
and the critical examination of the now numerous
recoveries of marked birds, have qualified many
theories of the past, strengthening or confirming
some, proving the futility of others.   Yet many
problems await solution, further light is needed.
That light will surely come ; science was never more
progressive than to-day.   Dr Eagle Clarke's "Studies
in Bird-Migration " and Dr A. Landsborough Thomson's " Problems of Bird-Migration " have been
invaluable in this revision.   T. A. C.

February 1929.

CONTENTS

LIST OF MAPS

NOTE TO THIRD EDITION

No attempt has been made to bring up to date the various designations of the authorities quoted, but throughout Mr W. Eagle Clarke should now read as Dr Eagle Clarke, whilst, to follow a general rule, titles should be omitted from the names of scientists of the past. Otto Herman, W. W. Cooke and others are, unfortunately, no longer with us.

THE MIGRATION OF BIRDS

CHAPTER I

MIGRATION OF BIRDS

MIGRATION is the act of changing an abode or resting place, the wandering or movement from one place to another, but technically the word is applied to the passage or movement of birds, fishes, insects and a few mammals between the localities inhabited at different periods of the year. The wandering of a nomadic tribe of men is migration; the mollusc, wandering from feeding ground to feeding ground in the bed of the ocean, migrates; the caterpillar migrates from branch to branch, even from leaf to leaf; the rat leaves the ship in which it has travelled and migrates to the granary; we pack our goods, hire a removing van and migrate to a new abode. The word migration thus applied may be literally correct but it fails to convey the generally accepted meaning, and the expression Bird Migration suggests periodical and regular movement, the passage as a rule between one country and another.

The popular application of a term does not do

A   1 away with the need of definition, especially as there are many complicated phases of migration. The migration of birds is as a rule between the breeding area or home and the winter quarters, but there are many migrants which never reach breeding quarters in spring, and many others which leave the regular breeding quarters or the place of residence in winter to perform a very real migration under peculiar stress of circumstances. Again the spasmodic movements of certain gregarious species, which at irregular intervals change their location in large numbers to take up their abode in another part of the range, is really migration, though it is now usually described as irruption, incursion or invasion.

Newton says (38) that bird migration is " most strangely and unaccountably confounded by many writers with the subject of Distribution," but the very act of the bird which extends its range, the first step in distribution, is migration. The histories of present-day distribution and migration are irrevocably interwoven ; as Mr P. A. Taverner remarks (51), " migration is a dispersal, and conversely, this dispersal, as it manifests itself, is migration," whilst distribution is the outcome of dispersal.

Broadly speaking, all birds migrate, though the length of the journey varies in different species, and in some cases in individuals of the same or closely allied species, from the merest change of elevation to a voyage almost as wide as the world itself. The sedentary red grouse nests on the moors, often less than 1000 feet above the sea, but " when snow-bright the moor expands " it feeds and resides in the cultivated valley, and as shown by the committee appointed to study grouse disease, not infrequently migrates from range to range across wide valleys. Many tropical birds, usually considered non-migratory, are subject to short movements, the origin and purpose of which is search for food and safe nesting places.

The knot breeds in countless numbers in Arctic Asia, Greenland and America, though only a few ornithologists have traced its home; it migrates to the Cape, New Zealand and Patagonia. The Arctic tern has a northern breeding range extending perhaps as far north as that of any bird, and it has been taken far to the south of South America in the Antarctic regions; if the thesis that the further north the bird goes in summer the further south it travels in winter is correct, as it can be proved to be with some species, some of these terns must annually travel about 22,000 miles (21). Between these extremes are an endless variety of distances travelled and methods of migration, with striking differences in the performances of individuals of the same species. Take one instance, a song thrush reared in a nest in our own garden. We may see and recognise this bird up to the middle of July, but what trained ornithologist can, yet, say with certainty where that bird will be by the end of the month or in three to four months time ? We know that all through the winter there are some song thrushes near the house, and that they are the birds which not only begin to sing early but actually nest with us ; we know too that before there is any marked immigration of northern thrushes there is a recorded emigration from our southern coasts, presumably of thrushes which have nested with us, beginning towards the end of July ; further we know that there is an autumn immigration of Scandinavian or other northern song thrushes, sub-specifically distinct to the expert eye, and some, small and dark, whose origin is by no means proved, as well as later emigrations of birds to the Continent or Ireland, both regular and occasioned by exceptional weather. Will our young July thrush remain in England or will it join one of these streams, and if so which ? We do not know yet. I repeat " yet," for the study of races, sub-species or local variations is commanding more and more attention ; the patient work of the " splitters," scorned by the old school of

" lumpers," will eventually solve many of the problems of to-day.

The ancients—a usefully ambiguous term—realised that birds migrated ; our immediate forefathers of two or three centuries ago realised that certain birds vanished in winter and wondered how ; and within modern times the phenomena of migration, the " mystery of mysteries," has been the subject of much study, speculation, and literary exposition. Indeed a full bibliography of migration would be a considerable volume. Even workers within the last few years have declared that certain phenomena were beyond human understanding, only to be explained by instinct, a word capable of most varied interpretation. In truth there is much to learn, much to which we must still answer—we do not know ; but the speculative theory of yesterday is now either myth or fact, and the theory of to-day may be proved true and add something to the data of which knowledge is built. The wildest speculations, based on slender locally ascertained facts or on no foundation whatever except the fertility of the brain, have been offered as solutions of the mysteries ; the literature of migration is a jumble of contradictions. John Legg, in 1780, said " In relating so many instances of unparalleled credulity, I confess I cannot suppress the irascible passion " (33), and Herr Otto Herman, only a few years ago, pointing out the ingenious dogmas "void of every firm foundation," says that "really it is a field in which every thinking ornithologist may create new theses to any extent and more or less incredible" (31).

Herr Herman's system of "ornithophænology," the accumulation of substantiated observations and facts, will not prove everything, but his work in Hungary, that of Dr Merriam and Mr Cooke in America, and of Mr W. Eagle Clarke in Britain, each aided by a numerous band of careful workers, are striking examples of what can be accomplished. Whatever errors future enlightenment may show in their conclusions their ascertained facts will remain positive knowledge ; theirs is not what Herr Herman himself described as "pretended authority."

In order to grasp the problems of migration it is necessary to get rid of the puerile and insular aspect of the subject, namely that migrants are merely those birds which come to us, like the swallow and cuckoo in the spring, and those, like the field-fare and brambling, which visit us in winter but are not with us in summer. The complication of the subject may be demonstrated by a rough classification of the migrants to be observed in the British Islands.

Arbitrary grouping of the members of an avifauna

is only for general convenience ; many species are represented in more than one group.

1. Permanent Residents : birds which remain in Britain all the year round. These are comparatively few in number, and largely consist of insular races of birds which perform regular and often long migration journeys in other parts of their range. Most, if not all, perform short migrations, in some cases only seasonal changes of altitude, spending summer on the hills and winter in the lowlands ; examples, the red grouse and dipper. Others, like the tits and creepers are nomadic and more or less gregarious in the colder months. Few appear to remain in the same locality at all seasons, but possibly some of our British robins and song thrushes, both sub-species of migratory Continental forms, may be non-migratory.

2. Summer Residents : birds which nest in our islands, leaving in autumn for countries to the south, and return in spring. In addition to the regular summer visitors, which all leave in autumn, this group includes a number of wagtails, pipits, finches and other birds which are represented in winter in our islands by a proportion which remain.

3. Winter Residents : birds which nest to the north or east of our islands and arrive in Britain in autumn, leaving in spring for their breeding area. With birds like the fieldfare, brambling and jack

snipe, which do not nest in Britain, must be included many (for example the robin, rook, song thrush and common snipe) which are also permanent residents.

4. Birds of Passage or Spring and Autumn Migrants : birds which neither nest with us nor normally remain for the winter, but merely use the British Islands as feeding and resting places on their journey between the northern breeding area and the southern or eastern winter quarters. This group is an especially difficult one, for in it must be included such birds as dunlins and curlews, which are represented as breeding species in Britain, and also a number of birds which apparently go no further south than our islands in winter, and others which, though not breeding, go no further north in summer. The actual status of these individual birds is uncertain. In this group too we have the Greenland wheatear, so closely allied to our familiar early migrant that, unless the bird can be measured, its identification is uncertain.

5. Irregular Migrants : birds which may be classed in other groups. Some of these are really winter residents, but their visits are so irregular that they may for convenience be classed with spasmodic or occasional invaders, such as Pallas's sand-grouse, which arrive at uncertain intervals in large numbers. Some of their number, during these irruptions, usually

breed and thus the bird becomes an irregular summer resident or even, for the time, a permanent resident.

6. Stragglers or Wanderers : birds whose occurrence in our islands is more or less accidental, due apparently to their having lost their way or to their ordinary wandering habits having taken them far from the normal range of their species. Some of the rarer petrels and other oceanic birds certainly pertain to this group, but our knowledge of the migration routes of others is still so slender that it is unwise to declare dogmatically that they are lost. Some too of the so-called stragglers may have been artificially or accidentally introduced ; many " records " prove on investigation to be the aimless wandering of escaped captive birds, whilst others are known to have been aided in their journey and carried out of their usual course when resting on shipboard.

When Mr Eagle Clarke was on the Kentish Knock Lightship, off the mouth of the Thames, he found that in autumn there were continuing practically simultaneously the following streams of migration. Immigration from the Continent to England from east to west, and from south-east to north-west, and passage along both lines ; emigration from north to south-south-west, and from north-west to south-east, with passage from north to south-south-west. Birds of the same species actually crossed paths, travelling in contrary directions (16).

The above grouping applies to the British avifauna, but a somewhat similar arrangement might be made of the birds of any particular area, large or small. The grouping of birds for the study of Geographical Distribution is of little consequence in connection with migration, but the mapping of the world into various ornithological rather than zoogeographical regions is of considerable importance, both for convenience in tracing the ranges of migrants, and in the discussion of the history of migration, which almost certainly began in the form of short wanderings from the centres of distribution. It is of comparatively small importance what boundaries we take for the various regions ; these depend largely upon the view of certain ornithologists as to which groups of birds shall be considered as typical of the regions in question. Sclater's six regions are perhaps the most universally used. They are as follows :—

1. Palæarctic, embracing the whole of Europe and northern Asia.

2. Ethiopian—Africa, Arabia, Madagascar and roughly half of the Atlantic and Indian Oceans.

3. Indian, including India, Further India, Southern China, the western portion of the Malay Archipelago and the Chinese Seas.

4. Australian, embracing Australia, New Guinea, New Zealand and the southern Pacific.

5. Nearctic, roughly America north of the Gulf of Mexico.

6. Neotropical, America south of the Gulf.

Newton suggested an alteration, a continuous northern region to be called the Holarctic Region, which embraces almost the whole of the Northern Hemisphere, and the division of the Australian into Australian and New Zealand Regions. Each of these southern regions is the winter home of some of the Holarctic birds, and it is a matter of dispute whether many of these originated in the northern or southern hemispheres. The value of these artificial divisions of the world is rather in the consideration of the conditions their varied climates and physical features present as attractions to birds in search of suitable nesting places and food supplies.

The study of Migration involves reference to the work of ornithologists of the past and present, the mass of contradictory literature already referred to, and we are repeatedly faced with the difficulty that some particular theory about the vexed questions of the cause or origin of migration, the height and speed at which birds travel, whether they do or do not follow routes, how they find their way, in what order they migrate, how and why they do or do not avoid dangers, or any similar problem, which seems to give finality so far as certain cases are concerned, is met by an absolute negation in other instances. The truth seems clear; more than one factor has influence on most birds, and different species in different places are influenced by different factors. Elliott Coues' sweeping statement, though I strongly disagree with the article in which it occurs, expresses much that is true. " Isepipteses and magnetic meridians, coast-lines and river channels, food-supply and sex-impulses, hunger and love, homing instincts and inherited or acquired memory, thermometer, barometer and hygrometer, may all be factors in the problem, good as far as they function ; but none of them, and not all such together, can satisfy the whole equation."

Some of the theses may be laws or rules, but there are no rules without exceptions, and these exceptions may become local rules. Laws regulating migration in one area, whether it be the great continent of America, the British Islands or the islet of Heligoland, may have little application in other parts of the world ; local evidence alone can never solve the great problems.

CHAPTER II

The question—What makes Birds Migrate ? or what causes them to remove from one zone to another at certain seasons, has been answered, no doubt to the satisfaction of the respondents, in many varied ways. Closely connected with the question of immediate impulse is the deeper, and less easy to prove problem as to how migration originated.

It has been dogmatically asserted repeatedly that birds invariably breed in the most northerly part of their range, and winter in the most southerly. Winter, when speaking of Holarctic birds, only applies to the season in the northern hemisphere ; the birds which pass south of the equator winter in summer. Whilst accepting this as a rule, two reservations must be made. First, that it only applies to birds of the northern hemisphere, and secondly that it is a rule with exceptions. It seems probable that the breeding area of some of the birds which reach the British Islands in autumn by the so-called east and west route is in more southerly latitudes than our islands, and certainly it seems

13

evident that the temperature of the winter refuge
has more effect upon the birds than its geographical
position.   Perhaps the statement that a bird always
nests in the coldest part of its range is more universally correct.   Even this may not be invariably
the habit, but in acknowledging it as a rule we
must clearly understand that this cold district is
resorted to at the period of the year when its temperature is at its highest.   There are certain birds
which breed in Australia and winter in Oceanic
islands where the temperature is cooler than in
their breeding area.

When considering the migration of birds which
summer in the extreme north or breed in the extreme
south—alas, but little is known about the migratory
habits of many southern breeders—it is comparatively simple to offer an explanation ; in the long
winter months this home, so desirable in the short
weeks of daylight, is dark, ice-bound, and foodless ;
it is wholly unsuited to the requirements of birds,
which, in spite of many assertions to the contrary,
have never been proved to hibernate, the only way
in which animals can survive for any lengthened
period when food supply is entirely cut off.

Birds are structurally provided with the means
of escaping from the disastrous effects of adverse
circumstances ; the power of flight, though not the
only way in which animals can migrate, is at the

root of the migration of birds. The advantages of the power of flight, to which also it owes its development, include the ability to avoid active and passive enemies, and to remove from one feeding ground to another undeterred by the barriers which restrict the terrestrial animal. A natural sequence of this ability to take advantage of aerial locomotion is the habit of wandering in search of food, more or less noticeable in all birds. The habit of wandering led to the discovery of feeding grounds and suitable nesting places ; where these nesting places, probably at first, only removed a short distance from the parents' nesting site, were suitable, dispersal and an extension of the distributional area or range of the species followed ; but where the feeding area was unsuited or not so well suited to the needs of the species, hereditary attachment to the original home and memory of the direction of this home, or even in some cases accidental wandering back to the more suitable locality, would originate a migration. Coupled with this are two important factors which would tend to make the habit periodical and regular both as regards time and locality. The memory of the bird, call it instinctive memory if we like, would limit the wanderings in search of food to a certain number of places where food was most abundantly found, and the passage between feeding area and breeding area become regular

journeys, at the seasons of the year when an increasing number of young birds in the breeding
area drove the overgrown population to seek food
further from the base, and again when the sexual
impulses urged the birds to seek secure nesting sites.
The other factor is the weeding-out influence of
mistaken effort, the natural selection which leads
to the survival of the fittest.   The young wanderer
which reached unsuitable lands must either wander
further or perish.   Judging by the juvenile mortality
amongst young birds the failures would be many,
and only the successful competitors would return to
leave progeny.

Great stress has been laid on the attachment of
birds to certain nesting sites, an undoubted fact,
and it has been argued that because, in some cases,
for hundreds of years certain sites have been
occupied by the same species, it is evident that
after the death of parents the young will return
to and occupy the home. This has even been
put forward as evidence that birds do not wander
in search of fresh nesting sites. The argument is
not sound. It is improbable that in most cases
both parents perish in the same year. Birds of
prey, and many of the cited instances of long tenancy
refer to raptorial birds, have a wonderful power of
finding a mate, male or female, to complete the
hatching and rearing of the young, when one of

a pair has been destroyed. The survivor of any pair might have the home attachment and by bringing a fresh mate create an attachment which would be passed on from mate to mate indefinitely. Again it must not be overlooked that certain sites present advantages to particular species which must be evident to all in search of those advantages ; it by no means follows that the occupiers of a nesting site are in any way related, except specifically, to those which occupied it in previous years.

The answer to the argument that birds do not seek fresh nesting places and thus extend their distributional area, is evident when we consider those species which, at the present time, are extending their range. Within the last few years, for instance, the turtle dove and tufted duck have begun to nest regularly in many parts of England in which they were entirely unknown twenty or thirty years ago. The starling has spread and in some parts is spreading still, and many other similar cases might be cited.

In this manner migration, as we know it to-day, may have originated, and as Mr P. A. Taverner expressed it, "however instinctive their habit may now be, there must have been a time when migrations were intelligent movements, intended to escape some danger or secure some advantage " (51).

B

Granting this, however, as the first cause, we are only on the threshold ; the question still remains unanswered, what actually impels the birds to seek fresh food supplies or to look for safe nesting places ? The natural answer, the cravings of nature and sexual impulses fails to give satisfaction in every case. Wanderings in search of food might lead in any direction, and probably did in the first place, but now birds in the main travel south in search of food and north in search of home, and many of them perform immense journeys, passing over or through lands which are capable of supporting a wealth of bird-life even in the winter months.

The majority of Arctic birds or those nesting in high latitudes leave before the great harvest of autumn fruits, and even our common swift begins to depart—for all do not go at once—towards the end of July, when insects are more abundant than at any other time of the year. Food supply has not failed when most birds start their journey in search of food ! Again in spring, when it is claimed that the powerful sexual impulses are sufficient reason to account for the northward journey, hosts of sexually immature birds and of others which are apparently mature but do not breed that spring, migrate northwards, some even arriving before the mature birds of their own species.

The earlier students of migration insisted that temperature was the sole cause of change of abode ; that the northern lands became unsuitable through their falling temperature, and that the birds deserted them for warmer climes, returning when the lands they wintered in became too hot. As a variant of this notion, which cannot be lightly cast aside, the suggestion was mooted that it was not cold but the lack of food during the cold months which drove them south, and that in the Tropics, where at one time it was thought that all migratory birds wintered, food was scarce during the months of extreme heat. Dr Wallace went further and stated that the incentive to northern migration was the inability to find sufficient soft bodied insects suitable for the nestlings in the Tropics during summer (**54**). Yet there are birds which do find food enough for their young, and some of them are insect eaters.

Seebohm, arguing with reason that the first home of the *Charadriidae*, was the Polar Basin (**44**), suggests that the desire for light originated the idea or the action, and though this was only applied by him to Arctic birds, others have striven to show that the longer hours of daylight would be an advantage to all birds, even though the difference of dark and light in the zone retired from and in that arrived at might be inconsiderable (**41**). Against this must be taken into consideration the fact that many waders and ducks, northern breeders, feed by night or day, according to the state of the tide. Light is not an absolute necessity to them.

The suggestion that migration owes its origin to the Glacial Epoch, " that supposed solution of so many difficulties," to quote Mr Gadow (28), has had many exponents. Some take for granted that the Polar Regions were the original home, the centre of dispersal, of all northern birds, and consequently that migration originated in the gradual pushing back of avian life as the ice gained more and more land each year. During the summer, the birds, urged by an irresistible love of home, travelled as far north as the ice allowed them, but gradually they were driven to nest further and further south until they found refuge in the un- glaciated parts of the earth. The individuals and the species, if not the whole families of birds, which failed to retreat, went the way of the " thousand types." On the retreat of the ice, the birds, impelled by a mysterious hereditary memory of home and of the good times enjoyed by their remote ancestors, for very very many generations must have been born under more or less sedentary conditions during the Ice Age, began the same pushing forward each year to the limits allowed them. In this case they travelled nearer and nearer to the original home instead of constantly being driven further from it.

Surely the question of original home, at anyrate of the home in pre-Glacial days, may be entirely left out of the question. No one can ever prove that this wonderful memory did or could exist. Post-Glacial dispersal northwards, and the foundation of migratory habits of advancing to the new food-producing areas, suitable also for the rearing of young, was doubtless a fact, but would have taken place in any case. The congestion due to the increased numbers driven to a restricted area, would involve a rebound outwards, and the uninhabited areas northward of the refuge would be the natural bourn towards which the birds would travel. The seasonal return of cold would drive them southwards in winter, and the periodical migration habit would thus be originated.

The intense love of home during the spread of glacial conditions would tend rather towards extinction than the formation of any new habits. The birds which possessed the greatest attachment to the particular district would be less likely to fly from adverse conditions, and the reduction of their numbers through the ordinary physiological changes in habit—reduction of the number of young produced, and possibly disinclination to pair—would inevitably end in extinction. The stronger the attachment to home the more likely the bird to remain to the bitter end, and if driven away by increasingly severe winters, to return and attempt to nest in the locality which had become unsuitable for nesting. The spread of glaciation would be gradual and so would be the annihilation of the species, but the end would be sure.

Birds which are cited as species which have shown this remarkable attachment to home, have disappeared before adverse circumstances — the great auk and the Labrador duck.

From what little we do know about the behaviour of our summer birds in their winter home, we may safely conclude that their habits are similar to those of winter visitors to Britain. Only in a few species are there two restricted areas, two abiding places or homes. The necessity of retaining a secure home for the young and the care of these young during their more helpless age keeps the individual birds within a certain area during the breeding season, but at all other times the bird is more or less of a wanderer. The variation, however, of the wanderings is remarkable. For instance the flocks of fieldfares, redwings, and some of the finches which come to winter in the British Islands wander continually from feeding ground to feeding ground, remaining in one place only so long as the food supply is plentiful. When there is a plentiful harvest of beech-mast, chaffinches and bramblings will linger near one clump or avenue of beeches for many weeks, but when, as often happens, the mast crop fails, they become nomadic, and pass from place to place in their hunt for food. They visit fields top-dressed with manure, glean the refuse of the harvest, frequent the farm-yards, and in early spring, visit the budding larches to prey upon their insect pests. On the other hand golden plovers and lapwings are remarkably local in their winter habits, and so long as the weather remains open will frequent the same fields throughout the winter. Severe weather, especially snow, which effectually closes their chance of obtaining food, at once drives them away. They will migrate to the unfrozen mud-flats of the coast, or to those parts of England, generally the south-west, and Ireland, where the climate is normally milder, or they will even leave our islands altogether under great stress.

The wandering habit, except during the breeding season, is confirmed in most birds, and experience shows that the same species of birds visit the same districts again and again when there is some particular food supply to attract them. Memory and experience guide them from place to place. This regular visitation of certain food bases, being of the greatest importance to birds which have a long period of travel or wandering before them, tends to originate the so-called route by which they travel. The fact that as a rule these stages are in consecutive steps southward is surely due to the fact that the temperature is falling in the north more rapidly than in the south. That they are not always due south is certain. The American golden plover, as Mr Wells W. Cooke so lucidly demonstrates, at first travels eastwards from its home in western Arctic America to the fruit-laden lands of Labrador and Nova Scotia, where it feeds for some time, stoking up for its long over-sea journey due south. Mr Cooke says, " It can also be said that food supplies *en route* have been the determining factor in the choice of one course in preference to another, and not the distance from one food base to the next. The location of plenty of suitable provender having been ascertained, the birds pay no attention to the length of the single flight required to reach it " (21). During the evolution of the route many bases would be found which were superior to others, and skipping and the gradual shortening of the journey from one to another would result. The final goal, the food base which in any weather or season provides the safe sufficiency of food, having been reached by the birds, this becomes the winter quarters. The return to this secure retreat each winter, instead of aimlessly wandering in search of a better, and thus the long-distance migratory habit is formed. Heredity tends to confirm this and it becomes an instinct.

Any observer may verify the assertion that birds regularly visit certain favourable food-bases by paying attention to the occurrences of birds of passage. The study of a county, for instance, shows that certain species show partiality for particular localities. Thus in Cheshire goldeneyes pass through every spring and autumn, and may be met with occasionally on any of the meres; but at Oakmere, in the Delamere district, one may be almost certain of seeing parties of this species any time during the periods of passage. Before the winter resident golden plovers have arrived in autumn and after they have departed in spring, the favourite fields are regularly visited by passing flocks. Chance may lead a casual wanderer to a good food-supplying spot, but the regularity of appearance suggests habit and memory. Very noticeable is the regularity of visits of passage waders to sewage-farms, both on spring and autumn migration. As the municipal sewage-farm, with its settling-tanks and withy-beds, is a comparatively recent introduction, these food-bases must have been discovered by chance. No hereditary knowledge could guide the birds.

A fact which supports the theory that birds ramble far in search of food in their winter quarters, is that in many species the winter range is more extensive than the breeding area. Thus Mr Cooke shows that the known breeding area of

the Pacific golden plover has an east and west
extension of some 1700 miles, but in winter it
ranges over an area with an east and west extension of about 10,000 miles. The scarlet tanager,
however, has a breeding range extending for some
1900 miles across eastern Canada and a winter
home in north-western South America of only
some 700 miles in extent.

The winter quarters, or the outermost limits of
the individual but not necessarily the specific range,
having been reached, the bird spends its time in
seeking food, remaining in one place if food is plentiful, or wandering, according to necessity or the habit
of the species. The assertion that some birds have
a second breeding season in their southern home
is either unsupported by any direct evidence or
is the result of a mistake in identification ; the
bird which has been found breeding has in
several instances been shown to be a southern
form or a related species of the one it was thought
to be.

As the northern spring approaches, the strongest
of all animal instincts, on which reproduction and
the very existence of the species depend, overcomes
all other desires, and the bird grows restless. The
hereditary instinct, the origin of which we have
endeavoured to show, urges the bird to seek the
breeding area which has by degrees become so far

removed from the winter quarters. The bird returns home.

But here is a serious difficulty urged by some writers as a powerful argument against the sexual impulse as the great factor in the return journey. Many of the birds which migrate northwards or homewards are sexually immature, and others of them are undoubtedly to be classed as "nonbreeders," which means that during that particular summer they will not be engaged in the work of reproduction; why, then, should young birds or nonbreeders migrate from the winter base. Possibly in the early days of migration only the mature birds did return; that we cannot state one way or the other. But it is reasonable to argue that once a regular migration habit has become not only confirmed by heredity but a very true advantage to the species, its influence will be felt by each and every individual. Again it is clear that the sexual impulses, in an undeveloped form, are appreciated by the adolescent, and in many animals by even the most juvenile. The play of all young animals is either an imitation or reflection of the search for food—the hunting instinct—or the love-making and sexual quarrels pertaining to reproduction, the pretended competition by the young for the favours of the opposite sex. They may play at and actually perform a migration which is so closely bound up

with the life of the species. That this impulse has
not always sufficient strength to force them to perform the whole journey is apparent from the fact
that many non-breeders, young or sexually mature,
on their northward journey through our islands or
along our coasts, never reach the breeding area ;
the food supply on the way attracts them more than
the memory of home ; they linger with us until the
breeding season is over and the return journey has
begun. Knots, sanderlings, turnstones and many
other waders may be seen on passage late in June,
and some remain on our mud-flats throughout the
summer ; in July the tide of migration has turned.

It has been suggested that some of the sexually
mature non-breeders may be actually enjoying their
winter during our summer ; in other words that they
have bred in southern breeding-stations whilst their
congeners wintered in the same zone. This means
a double breeding - area for certain species — a
possible explanation, but one hardly supported by
known facts. When a bird had so cosmopolitan a
range that in the course of its dispersal its breeding
areas were separated, we almost invariably find that
the birds inhabiting these two areas are distinguish-
able geographical forms or sub-species. Mr W. H.
Hudson, in his " Naturalist in La Plata," refers to
the godwit, *Limosa haemastica*, which spends the
southern summer in La Plata and breeds in the

north, and to birds of the same species which winter in La Plata, arriving from supposed breeding places to the south when the northern birds leave. Captain R. Crawshay, author of " The Birds of Tierra del Fuego," found it in this little known land, but speaks somewhat doubtfully of its identity ; we shall probably learn that the southern form is sub-specifically distinct from the northern. There are other wide-ranging waders which are suspected of having a southern nesting area, but we still await proof.

The lack of sufficient or suitable food in the winter home during our northern summer may also cause the exodus, but this is a difficult point to prove when it is remembered that the winter home of every bird is not the parched tropical land or the waterless desert. From some zones removal must be a necessity, but in others there is food for all, so far as man can tell.

Dr J. A. Allen, a severe but discriminating critic of migration theorists, says—" Migration is the only manner in which a zoological vacuum in a country whose life-supporting capacity is a regular fluctuating quantity, can be filled by non-hibernating animals " (51). When in the early days of migration this periodically-supplied northern zoological vacuum was filled to overflowing by the increased numbers of avian inhabitants at the close of the breeding season, the natural food supply would be taxed to its limits ; the falling temperature drove some and finally all to seek food further south, and their short migration to lands already filled with old and young birds, caused pressure and overcrowding further south.   Further outward and usually southward movement was necessary and the zone of stress was gradually extended, though probably in those early days no particular species took long passages. The winter passed and the vacuum was again provided, and the rebound to fill it would create a slackening force all along the line ; birds would spread from congested districts so soon as food supplying areas opened to receive them.

Mr Taverner, arguing on these lines (**51**), shows that competition would be originated in areas containing the earliest breeders, and be severest in the most productive districts.   Weaker and later breeders would be driven out or prevented from colonizing by the stronger and earlier species, and the evicted ones would encroach on others, forcing them in turn to trespass on a wider circle of species. He then argues how the gradual recession of the glacial ice would increase the possible northward breeding area, and cause longer migration, and that this migration would delay breeding and conversely delayed breeding would assist the evolution of migration.

But the lengthening of the journey might surely

be occasioned in another way, and the evolution of migration assisted apart from any glacial influences. Each successive increase of the length of the journey taken by the stronger and more go-ahead individuals, leading them in advance of the bulk of southward moving and competing birds, would be a distinct advantage to the individual and consequently to the species. The pioneer would arrive, like the slower movers, in a land already peopled with an avian population, but it would not have its own fellows to add to the stress of competition ; it would be ahead of the greatest struggle. So the fittest would mould for the species the most suitable journey both in distance and route, and the laggards would gradually fall out of the competition.

Dr Wallace, without destroying these arguments, has shown that the survival of the fittest has a powerful influence. Those birds which do not leave the breeding area at the proper season will suffer and ultimately become extinct, and the same will happen to those which fail to leave the winter quarters when it would be a distinct advantage to the species to move into lands better suited for reproduction.

It has been put forward as a serious objection to many arguments that migration, instead of being advantageous to birds, is a danger to the race ; that the perils of the journey are greater than those occasioned by more sedentary habits. It has even been suggested that migration is a habit specially created to thin down the surplus bird population. Dr W. K. Brooks, however, puts this idea, which is not entirely devoid of truths, in rather a different way. " Adaptations of nature are primarily for the good of the species—beneficial to individuals only so far as these individuals are essential to the welfare of the species " (9). The destruction of overabundant young, the thinning down of superfluous numbers, may be an economic advantage. It is one thing to say that migration has been caused to kill off a surplus, and another to show that, once a habit has been originated and become an advantage, it will be conducive to a greater prolificness, and that the natural sequence of an increased birthrate, when food supply and other conditions remain unchanged, must be an increased mortality. Thus the perils of migration may become a boon to the species.

The theories of C. L. Brehm (7) and Marek that birds are living barometers, foretelling by intuition the changes of barometric pressure, may be dismissed as purely speculative. That birds begin their journeys during particular barometric conditions is certain, but what they know of forthcoming weather conditions is guess-work.

CHAPTER III

ROUTES

THE migrating bird, when passing between the breeding home and the winter quarters, travels by what is termed its Route. The definition of the route has caused more controversy than perhaps any other incident of migration; the chief point at issue is whether the bird uses a particular high road, along which all its fellows from the same area travel, or if all birds move in what has been called a " Broad Front." Ornithologists have been, and to some extent still are, divided into two camps, one upholding defined routes and the other the extended or broad front movement.

After all the difference is merely one of degree. Even the widest notion of the broad front, that of Gätke, who insisted, as dogmatically as he did on most points, that the width or breadth of the migrating host corresponded with the extent of the breeding range (29), is of a route, bounded on the one hand by the northern or eastern and on the other by the southern or western lines of latitude or longitude which marked the limits of the range. The idea of a route may be narrowed down to the extent of a wide river valley, or to a fly-line represented on a map by a ruled line, which passes over certain ascertained places. The absurdity of Gätke's arguments are proved by the study of his truly remarkable book. According to him the island of Heligoland was only remarkable in that it possessed an observer, himself, who saw marvels unobserved elsewhere, though the same number of birds were every year passing over any particular spot in an area which, for many species, must have been many degrees in extent.

Had not so much weight been placed upon, and so many arguments based on Gätke's extraordinary statements by, unfortunately, many of our leading British ornithologists, his theories might have been ignored. Unfortunately he is looked upon as an authority, even an oracle, whereas, as Dr Allen pointed out, on many points which he treats with great positiveness his knowledge is obviously as limited as the little field which was the scene of his life-long labours (2). Glibly he tells of hooded crows " in never-ending swarms of hundreds of thousands " passing across and for many miles on either side of the island ; of " every square foot of the island " teeming with goldcrests, and of " dark autumn nights " when " the sky is often completely obscured " by the migrants, which pass thousands of feet overhead. How did he observe the obscured sky ? Indeed he again and again declares that migration passes unseen yet calculates the numbers observed on the darkest nights ; the illumination of the lighthouse could not be sufficient to enable him to even guess at the numbers he mentions. After stating that " the whole vault of heaven was literally filled to a height of several thousand feet with these visitors from the regions of the far North," and that a certain east to west passage extended from the Faroes to Hanover, he concludes that " the view—that migrants follow the direction of ocean coasts, the drainage areas of rivers, or depressions of valleys as fixed routes of migration can hardly be maintained."

As emphatically he maintains that most observable migration over Heligoland is due east to west or west to east, though the birdstuffer Aeuckens, who supplied him with much of his information, told Seebohm that it was north-east in spring and south-west in autumn (45). Is it not perfectly evident that the geographical position of Heligoland makes it a convenient resting place for large numbers of migrants, for it is certainly true that large numbers are observed there, which pass southward and westward along the Baltic, crossing Schleswig-Holstein and the mouth of the Elbe, or coast south along Denmark, and cross the Elbe diagonally, *en route* for the Dutch and French coasts and to a lesser extent the south-east coasts of Britain ?

Coasting undoubtedly exists ; birds, day migrants especially, may be observed following coastlines in steady flight, though a mile or less inland no passage is visible. On the North Norfolk coast I have seen little parties of swallows passing along the shore in spring, coasting slowly but steadily from east to west. All day long and almost every day for more than a week this steady flight was continued, though I never saw any passing within sight more than a few yards out at sea, nor any at all more than a few hundred yards inland. Evidence which cannot be refuted shows that this habit of coasting is general, though a deeply indented bay, an estuary or strait, is usually crossed, and by no means always at the narrowest point. The same careful observations prove that both narrow and wide river valleys are followed by migrating birds in greater numbers than are ever observed passing beyond the limits of these valleys.

Seebohm's experience in Siberia led him to doubt the existence of routes, but his later studies of migration in autumn at Arcachon and in spring at Biarritz, caused him to modify his ideas. He found a gentle but continuous stream of migrants following the coast of the Bay of Biscay, arriving from over the Pyrenees on their northward journey, but moving " only within a mile or two of the coast." He contrasts island and coastwise migration ; in the latter the travellers can rest at night or take short journeys during bad weather, but in the former they must await favourable conditions before attempting a perilous passage (45).

On the other hand many birds undoubtedly pass over inland localities independently of any river valley or mountain range which might indicate a route. Even such typical coast-lovers as the maritime waders constantly cross or pass through inland England. They are heard at night, or met with resting or feeding on inland waters, or their bodies are found when, on a dark night, they have collided with telegraph or telephone wires.

So long ago as 1886 Mr W. Brewster maintained that the breadth of the fly-line varied according to the character of the country which was being crossed. The migrating column, he said, might be hundreds of miles in length, " a continuous but straggling army," which only became a " solid stream " when travelling through some narrow pass (8). This solid stream or army passage is, however, seldom observed when the birds are crossing continents, especially if they are traversing a wide area in which food is equally plentiful for miles on either side of the direction of flight. The consolidation of their numbers appears only to take place when, either on account of the indifferent food-supply or of unsuitable weather conditions, the speed is accelerated.

In America Mr Cooke proves that the Mississippi Valley is undoubtedly utilised as a fly-line by a large number of species, but by no means all, and his evidence, though proving the use of routes, is that these are seldom constricted pathways but broad areas crossed in a generally coincident direction by the birds which make use of them. This main fly-line is however formed in America as in other places by the convergence of subsidiary streams, and it is these tributaries, as Herr Herman points out, which have in many instances led to error ; they have been mistaken for main routes. The main route may be compared to the trunk of a tree, the birds following the roots from the area in which they have been nesting or wintering, and at the end of the journey splitting off in various directions, like the branches, to their temporary winter or summer homes.

The contrast in the method of travelling of different species or of the same species under different conditions, may be realised by taking two examples. Firstly, Mr Eagle Clarke's experiences at the Eddystone and Kentish Knock Lightship, when birds passed during the daytime at varying elevations, sometimes close to the waves, in twos or threes or scores, and at night in large numbers. The other is an observation of a " bird wave " by Mr P. Cox, during a snow storm in 1885 at Newcastle, New Brunswick. The birds passed eastward in a column about twenty-five yards wide, some just above the trees, others hardly visible, but the bulk in a massed column directly over the margin of the shore, and not over the river or meadow on either side. The movement was continuous for about two hours.

Dr I. A. Palmén was the great upholder of routes in the Old World, but his routes were largely speculative ; they were founded on a considerable knowledge of migratory birds, but not sufficient to cover the vast area mapped out (**39**). Until a very large band of workers, working on similar lines all the world over, accumulate a sufficient mass of evidence as to which birds do or do not pass their various stations, with the times at which they appear, accurate knowledge of the routes of birds is impossible.

Von Middendorf collected statistics of the passage of birds in the Russian Empire, and by reckoning the average date of arrival of a few species at certain points of observation, worked out a number of curves or lines which he calls " isepipteses," or lines of simultaneous arrival (**35**). The result was, according to his argument, a general convergence northwards ; the birds passing through Central Siberia travelled roughly in spring from south to north, in Eastern Siberia from south-east to north-west, and in Europe from south-west to north-east ; they converged, in fact, upon the Taimyr Peninsula. This to some extent is doubtless true, but Middendorf goes on to prove that the magnetic pole is situated in this Peninsula and that the birds are drawn thither by magnetic influence, " in spite of wind, weather, night or cloud." He calls them " sailors of the air," possessed of an internal magnetic influence. He supports his argument by the statement that there is a similar convergence in North America towards the magnetic pole of the western hemisphere.

But all birds do not go in the direction of the magnetic poles, and many of those which do, stop short at suitable breeding places long before they have travelled so far north. The Taimyr Peninsula, a vast area in the extreme north of Siberia, is each spring a " zoological vacuum "; towards this desirable spot migrants will stream.

Herr Otto Herman cleverly shows the absurdity of many of the reputed routes by cartography ; his map is crossed in all directions by the routes upheld by various theorists. Birds could not possibly follow all the directions " which authors invented for them," most of which he adds are founded on mere supposition (31). Dr Palmén, he shows, usually managed to avoid districts where there were no observers, but Mr Dixon and M. Quinet made their routes follow rivers and coast lines, whether there was evidence to support this idea or not.

Only to a certain extent can it be safely contended that the present route of a species is an indication of its earlier journeys, or that the direction of original dispersal is recapitulated in the present line of migration. Heredity, experience, and imitation would certainly tend to preserve and confirm the general direction ; the shortest and easiest passage from food - base to food - base would become an hereditary route, unless circumstances arose which caused a change. Mr Cooke shows how there has probably been evolution of the route as well as of everything else concerned with a mutable animal. The fly-line across an arm of the sea may be lengthened if this lengthening means a corresponding advantage in reaching the desired haven. Thus the birds which now cross the Gulf of Mexico at its widest part, at one time probably coasted round the Gulf, as many do still, by the land-bridge of Mexico and Central America. The gradual straightening of this curve would shorten the journey both in time and distance, though lengthening the actual single flight across a portion of the sea. We can imagine a bird arriving in autumn at the mouth of the Mississippi, at first passing from Louisiana to Mexico, so as to save the time of travel through Texas. Generations later the shortening of the journey, through lengthening of the short cut, would lead the birds to Vera Cruz and later still to Yucatan. It may be questioned, what object could the birds have in risking an oversea voyage, away from chance of food and hope of rest, when the land-bridge remained open for them ? Each individual or group of individuals which arrived at any particular place a little in advance of the migrating multitudes of its own species, or others which fed upon the same kind of food, would certainly gain advantage, and would be the most likely to develop strong flight and the power of endurance in its descendants ; it would indeed be a winner in life's race.

Great weight has been placed upon the use of land-bridges and the hereditary habit of crossing seas where these land-bridges once existed but have been submerged during the great geological changes in the earth's surface. Many have insisted that wherever migrants cross the sea they do so along submerged coast-lines or over submerged land-bridges, arguing that the gradual evolution which has made the advantageous adoption of a habit of migration possible was unable to eliminate the hereditary tendency to follow the exact route by which their ancestors passed from place to place. That there have been considerable alterations in coast-lines and in the general distribution of land

and water since the time when birds began to be migratory is indeed probable, but unless crossing the sea means a distinct advantage it implies the retention of a habit which would not only be useless but might be a positive danger to the species.

In the Gulf of Mexico and the Caribbean Sea there is evidence of perhaps the most recent land-bridge in the chain of islands from Florida to Venezuela, collectively known as the West Indies. Although vast numbers of North American birds winter in South America only a few of the species which annually pass from one continent to the other make use of this comparatively easy passage. One might naturally conclude that the final severance of England from the Continent was in the neighbourhood of the Straits of Dover, yet this short passage is only used by a comparatively small number of our migrants.

Mr Dixon indeed argued that there is no greater barrier to migration than even a narrow arm of the sea (26). He refers to many Continental species which are common breeders in France but are unknown as nesting species in the British Islands, and others which are found in England but not in Ireland. But this is surely but an incident of distribution ; the narrow strait or even river may for a time mark the limit of expansion of a species, just as at the present time the westward and northward unseen barrier prevents the range of the nightingale from spreading to districts apparently well suited for its home, and until recently the turtle dove and great crested grebe were checked in their northward advance.

In the evolution of some routes land-bridges certainly appear to have played their part, but once those bridges have ceased to influence direction the shortening of the time occupied by the lengthening of the single oversea flight is only a question of generations when an advantage to a species is to be gained.

This subject will be further dealt with in connection with the actual passages performed by certain birds.

The study of migration, based on observations at our lighthouses and lightships, shows that even in the comparatively small area of the British Islands there are certain routes followed with regularity. The birds which pass along our western coasts of England and Wales do not as a rule follow the shores round Cardigan Bay or along the eastward tidal scoop of the Irish Sea towards the coasts of Lancashire ; the main body passes from Pembroke to the Lleyn Peninsula, and thence to Anglesey and the Isle of Man, on its way to the southern Scottish shores.

A source of possible error in the method of deduction from these results must be taken into consideration. The observations at lightships and lighthouses are mostly made when untoward circumstances bring the birds within range of vision, and on dark and foggy nights cause them to strike the light in great numbers. What is their normal course when no great migration wave or " rush " is observed ? Are the few passing stragglers noted all that go by this route in fair weather ? The same uncertainty must be applied to the observation of passing birds in inland localities. The immense numbers which do pass is shown by the observation of large movements, when as occasionally happens some check to normal migration leads an army of birds to a dangerously low altitude, or when high winds hold up a portion of the host on our coasts ; but even these multitudes must be small compared with the millions of birds which annually pass from zone to zone unseen. The few or many birds we meet with, either on the coast or inland, resting on passage, may represent a lost or wandering party of stragglers or weaklings from a vast army which has passed over ; they may or may not be on the route or course normally followed by the majority. The cartography of bird migration is a study in itself.

Mr Abel Chapman, describing his experiences in the Mediterranean, says—" For forty hours we were passing across (or beneath) the lines of an army of migrants—say 500 miles in width ; yet not a sign did we see, save only the wreckage—the feeble that fell out by the way." On April 10th a sudden bitter northerly gale sprang up, and two hours later the steamer was the goal of hundreds of birds, no longer able to face the adverse wind. These were blue-headed wagtails, swallows, martins, pipits, wheat-ears, nightjars, and lesser kestrels. He thinks that the strong ones may have passed on but all the others perished (12).

My belief that broad-front migration is more general than the use of restricted routes is strengthened by recent observations. Even the Nile Valley, used by vast numbers of migrants, is not the only avian road to and from Central and South African winter quarters. Admiral Lynes, Mr R. E. Moreau and others have shown that not only do very many birds cross the Mediterranean direct to the North African coast, but that they cross the Sahara. That large numbers are met with in oases, but are not seen crossing foodless areas, suggests that they travel rapidly and at a consider-able height wherever there are no temptations to halt for rest or food. At a comparatively low altitude they would escape notice where few observers search the sky for passing birds. Nor is it necessary to wonder how Siberian breeders reach Indian winter quarters, since we now know that even the higher Himalayan passes can be crossed.

CHAPTER IV

THE HEIGHT AND SPEED OF MIGRATION FLIGHT

IN the last chapter reference was made to the great height at which birds may fly on migration. Certain species, even comparatively weak-winged ones, appear normally to fly high, whilst others, often birds with pointed wings and great aerial powers, usually proceed at low elevations ; but there is still much conjecture as to the actual altitude reached by any migrants.

Gätke was of opinion that we do not see much of real migration, which is certainly correct, but there is no reason for his statement that it normally takes place at anything like the altitudes he mentions, 30,000 feet or more, or that birds at the time of migration undergo physiological changes which enable them to fly at immense heights and speeds and to see clearly in the dark (29). Nor need much weight be placed on the speculation of Lucanus, who contends that the height of travel must be less than 1000 metres, for above that elevation aeronautical observations show that perspective lessens. There are actual observations which, though liable to a margin of error, are proof of migratory flight at very high altitudes.

Most of these observations were at first made by accident ; birds were seen through astronomical telescopes passing across the face of the moon or sun ; but recently this method of observation and several ingenious plans of measuring height and speed have been made use of expressly to study migration. W. E. D. Scott in 1880 thought that he recognised two species which passed across his field of vision at under a mile above the earth (**43**). In 1888 Mr F. M. Chapman calculated that birds passing at distances varying from one to five miles were at altitudes between 600 to 1000, and 3000 to 15,000 feet. " A number of birds were seen flying upwards, crossing the moon, therefore, diagonally, these evidently being birds which had arisen in our immediate neighbourhood, and were seeking the proper elevation at which to continue their flight," but the direction of most birds was parallel to the earth's surface and southerly. The average height was certainly far above the inferior limit (**13**).

Dr F. W. Very, whose original observations were recorded in *Science* (**56**), told me how he calculated the height and speed of birds by comparing them with lunar features. He obtained satisfactory results of 1643 and 1960 feet altitude, and speeds of 67 m.p.h. and 134 m.p.h. at 2000 feet. This last he concluded was due to the bird, travelling at normal rate, having risen into a rapidly moving air-current.

Many other interesting calculations have been made, but within recent years aviators have recorded actual observations of birds seen at great altitudes. Capt. C. Ingram's and Col. R. Meinertzhagen's papers in the *Ibis* for 1919 and 1920 prove that many birds can and do fly at great heights. Lapwings, for instance, were noted at varying heights between 2000 and 8500 feet, geese at 9000, small passerines of linnet size at 10,000 and 12,000, and rooks at 11,000. In addition I have information from friends of golden plovers at 6000, lapwings at 6500, starlings at 3000, and swallows at 10,000. There are numerous records above 3000 feet. Further, we have M. von Burg's actually thousands of observations of migration across the Alps, over crests and peaks as well as cols and passes. Wollaston's notes of redstarts, Richard's pipits, Temminck's stints, godwits and curlews at from 17,000 to 20,000 feet on the Himalayas, support Mr F. Ludlow's opinion "that birds . . . do cross and recross the Himalayan range in more or less direct lines." He found ducks, pintail snipe and other passage birds on lakes at 14,800 feet.

Mr Chapman's remarks about the upward flight of some of the birds are enlightening, for when birds start on oversea journeys they frequently ascend to a great height. Dr Allen and others think that the ascent is to increase the visible distance, but it may

D

also be to reach a zone or stratum of atmosphere in which flight may be more easily accomplished. Robert Service's account of the departure of migrants from the Solway shores, gives suggestion of high flight. They arrive often one by one and " seem to drop literally from the clouds," but when they actually departed it was easy to see their method. They " fly upwards and onwards, then they hesitate, fly sideways once or twice, again attempt an upward and onward flight, hesitate again, and down they come once more to earth." After repeating this manœuvre several times, " away they go over the sea." One morning he counted sixty blackbirds in one hedge, and others kept arriving, but, however closely he watched, he failed to see whence they came. " They came down from the upper air, becoming suddenly visible, sometimes three at a time." " I saw about a dozen birds thus drop into view, but I quite failed to see any indication of the point of the compass from whence they had come " (46).

Gätke frequently mentions birds raining down from the sky, appearing first as mere specks, and dropping vertically to the island, and others when departing " with breasts directed upwards and rapid powerful strokes of the wings, fly almost perpendicularly upwards."

On May 24th, 1911, I watched the departure of a spoonbill from Easton Broad on the Suffolk coast.

The bird rose and soared in ever-widening circles until it was a mere speck, even when seen through powerful prismatic glasses ; the great stretch of its wings alone enabled me to watch it for so long. When at a great height—I will not guess what elevation—it ceased its circling flight and made straight for the north.

In October I saw several flocks of redwings leave the Spurn. They rose to a great height before directing their flight southwards, although the Lincolnshire coast was plainly visible.

Great differences of opinion have been expressed about the speed of migrating birds, and here again Gätke's estimates, on account of the weakness of his arguments and his immense presumption, cannot be seriously considered. There are but few measured speeds, and most of these, except perhaps the ducks and geese referred to already, are of birds travelling at low elevations.

Many birds, especially day-migrating swallows, hooded crows and other birds, frequently travel at slight elevations ; it is not unusual to see birds at sea flying a few feet only above the waves. Mr W. Eagle Clarke, whose systematic observations demand the profoundest respect, again and again urges that the direction of the wind has little effect upon migration, but that the force of the wind may make migration impossible. At the Eddystone, where he

spent a month in the autumn of 1901, he noticed
birds passing at heights varying from 20 to 200 feet,
all flying southwards.   He concluded that " the
wind is certainly the main factor in migration
meteorology—I am convinced that the *direction* of
the wind is, in itself, of no moment to the emigrants,
for they flitted across the Channel southwards with
winds from all quarters " (16).   When the velocity
of the wind, however, was above 28 miles per hour
(a fresh breeze, force 5 on the Beaufort scale), no
migration was observed.

Allowing this, and also taking into consideration
the undoubted fact that birds are frequently held
up by strong winds on the shore before starting
oversea journeys, is there any proof that they do
not actually avail themselves of fairly steady strong
winds when they reach the upper air ?

Mr F. J. Stubbs makes some useful suggestions (50).
He points out that Gätke and others imagine that a
bird flying with the wind would suffer inconvenience
through the wind ruffling up its feathers.   It is surely
evident that a bird supported in a moving medium
could not progress at any slower rate than that
medium ; the bird is not flying on one stratum of
air with another moving with a different speed
immediately above it ; it is actually in a current,
not on it.   If the bird flies at 20 miles an hour, and
the wind or moving air is progressing at 10 miles an

hour, the bird will cover a distance of 30 miles in one hour, though the force exerted by the bird is the force necessary for 20 miles in a calm. Conversely, if a 10-mile air current meets it, it will unconsciously be carried only 10 miles. If the speed of the bird is the same as the opposing force of the wind, it will remain stationary ; I have seen ducks in a blizzard rise head to wind and fly rapidly, making no progress but maintaining their position over the water, to which they dropped again when the storm passed. Some black-headed gulls on the same water did not attempt this manœuvre, and in a few seconds had vanished down wind. The swimmer, in a swiftly-flowing river, may hold himself in position so long as he can swim at the same rate as the stream he is contending with, but he cannot make headway if the speed of the water exceeds his. He may, however swiftly the stream moves, swim in any direction, but his actual progress will be down stream ; if he aims to swim directly across, his real course will be diagonal.

The fact that birds fly in any direction in a wind, and when at low elevation pay little heed to the direction of the wind, when the breeze is light, simply means that they can fly faster than the medium they are in ; if the medium travels faster than they do, they will be carried in it to their advantage or disadvantage.

We have learnt much about wind in the upper air in the last few years. At varying altitudes it may differ in both direction and force. Usually the force increases at from one to ten thousand feet, whilst the direction may be exactly reversed at a high altitude.

Mr Abel Chapman, in " Wild Norway," makes this pertinent remark—" Except by aid derived from the operation of physical laws, the nature and extent of which are unknown to me, and by taking advantage of ' Trade-wind ' circulation in the upper air, I believe that migration is impossible for short-winged forms of sedentary habits—but that aid, and those advantages, may facilitate, and perhaps vastly accelerate, a process which is otherwise impossible."

In " Bird Life of the Borders " he goes further. " Birds are warmer-blooded than ourselves or other mammalia, and are capable of sustaining life in rarified atmospheres where these could not. By a simple mechanical ascent, they can reach, within a league or two, regions and conditions quite beyond human knowledge : where, selecting favouring air-strata, they may be able to rest without exertion ; or find meteorological or atmospheric forces that mitigate or abolish the labours of ordinary flight, or possibly assist their progress. . . . It is in the upper regions of open space where, I suggest, the final clue will be found " (12).   I think the clue is found.

A warbler flying leisurely, say at 10 miles per hour, in a current of air which was travelling at 20 or more miles an hour, could accomplish the journey across the North Sea—say 300 miles, in ten hours. Allowing much higher rates of speed for strong-winged species, and greater force of wind, some of the marvellous distances covered by migrating birds cease to be mysterious. Prof. J. Stebbins and Mr E. A. Fath made careful calculations from observations with the telescope, and found that birds passed at rates varying from 80 to 130 miles per hour, and these were the minimums, for if the birds were not flying absolutely at right angles with the line of observation, they must have travelled a greater distance in the time occupied between their passage of the observation points (47).

At the end of December 1927 many lapwings crossed the Atlantic to Newfoundland. I shall refer to this later. Mr H. F. Witherby, aided by information supplied by the Meteorological Office of the Air Ministry, estimated that in or on the 55 miles per hour gale the birds crossed in twenty-two or twenty-four hours, at an average rate of 100 miles per hour. Col. Meinertzhagen's estimate of the normal aerial pace of lapwings is 45 miles per hour.

CHAPTER V

ORIENTATION AND ROUTE FINDING

THE question, How do birds find their way? is answered by many ingenious and often purely speculative theories, some of which have been already referred to in connection with the points discussed.

Each theory, though it may apparently explain certain phases of migration, can be answered by some exceptional difficulty which makes it fail as a full explanation ; we are driven to the conclusion that birds possess a sense of direction, which is often, very incorrectly, called Orientation. Biologically this term does not imply any connection with the East, but is simply used to describe the power of finding the way back to a certain base, or of returning home. It is a power or sense which undoubtedly exists in many vertebrate animals and in some invertebrates, though it is hard, in many cases, to separate or distinguish it from memory and impression gained through eyesight. Mr John Burroughs, one of the strongest opponents of what he calls the "Sentimental School of Nature Study," gives in his "Ways of Nature" a striking instance of this faculty which may serve as an example, though the cases of dogs and cats amongst domestic animals, and of many wild creatures finding their way, could fill many volumes. His son brought a drake home in a bag from a farm 2 miles away, and shut it up in a barn with two ducks for a day and a night. As soon as it was released it turned its head homewards, but for three or four days its efforts were frustrated. Then Mr Burroughs decided to see what the drake would do, and to go with it to give it "fair play." The "homesick mallard started up through the currant patch, then through the vineyard towards the highway which he had never seen," and Mr Burroughs followed 50 yards behind. A dog scared the bird and turned it up a lane, but after a detour it reached the road again; it stopped to bathe in a roadside pool, then started off again refreshed. A lane, leading in the right direction off the main road, puzzled it, and it took a wrong turning, but, discovering its mistake, made for the road again, but not by actually retracing its steps. The false move seemed to put it out, for after hesitating at the next and right turning, it actually overshot the mark. Its companion, unable to spare time to continue the experiment, then headed it back, and when it reached the turning again it seemed to recognise something familiar and raced home with evident signs of joy.

The duck was a domesticated bird, but the incident is not without interest; the homing faculty was clearly exhibited, but it was not infallible; the bird made a mistake. So, inexperienced young birds, travelling instinctively by orientation may, and do, make mistakes.

Human beings, in varying degree, possess a sense of direction, and some a wonderful power of finding their way in strange places; it is most marked amongst those men we choose to call uncivilised, who, indeed, live in closer touch with nature than those of us who depend so much on compass, map, road, train and tram; we, as path-finders, are degenerate. Middendorf marvelled at the powers of the Samoyeds, but when he questioned them was met by blank surprise, and the cross-question—" How does the little Arctic fox find its way aright on the great Tundra ? "

In addition to this instinctive power, the bird has eyes and brain. We can afford to put aside as purely speculative Middendorf's suggestion that the bird is impelled or dragged by magnetic force, but we cannot deny that it uses its eyes and that it has a wonderful memory; its second journey will be easier than the first, for it will recognise landmarks, just as the drake recognised something familiar when it neared home. Sight, however, cannot be always necessary, for, at the Kentish

Knock Lightship, Mr Eagle Clarke noticed that birds flew so low along the water that they could not possibly have seen their way.

It is purely speculative to say that the young bird which travels alone, for it seems certain that many young do travel unguided on their first journey, has an inherited memory of the actual route to be followed, but that it has an hereditary sense of direction, or an hereditary impulse to travel in a certain direction, is quite another matter. The sea, to the young bird, may be a barrier ; it may wander coastwise and be lost, or, if this is the best way, find itself at the desired haven. If the shortest and quickest way is across the ocean, the young bird may brave the perils and succeed, or on this trackless waste it may wander till it sinks to the waves and be added to the long list of failures.

Certain species summer in Greenland either in the same areas or in areas comparatively near to one another ; some of these travel in autumn southeast, and winter in Europe or Africa, and others go south-west into the States or South America. Most of these are distinctly eastern or western forms, but occasionally American birds are met with in Europe, and European in America. Probably at the start these stragglers joined the wrong band, and travelled for company and unconsciously by the wrong route. Birds drifted to leeward may find companions of quite an alien tribe; others, wind-swept in this way, may travel alone to new lands. A few pelagic South Pacific and South Atlantic petrels, other than the great and sooty shearwaters and Wilson's petrel, which come annually to northern waters in their winter (our summer), have reached Europe from time to time, but theirs is an error of too wide nomadic wandering. Probably many more come than are observed; it is the inexperienced or weaklings, out of normal zones, which come ashore starved and exhausted.

Mr C. Dixon emphatically declares that each party of young birds is accompanied by one or more older ones to act as guides—"The many winter'd crow that leads the clanging rookery home." This is as emphatically denied by other observers. Probably there is no regular rule for many species, as there certainly is not for swallows, in which the old and young birds can be easily distinguished. I have seen bands trailing south in autumn in which all the birds were immature, and others in which a number of young were accompanied by a few mature birds, though, certainly, these old birds did not appear to lead. Another assertion, that the old and young do not even travel by the same route rather supports the idea that the young birds find their way simply by a sense of direction, a sense liable to blunder, whilst the old birds travel by the perfected or best route which their experience has taught them. Martorelli wisely asserts that orientation is not infallible, but develops with age.

Mach-Bruer attempted to localise this sense of direction in the semicircular canals of the ear, which are so highly developed in birds, but Dr Allen contends that the theory has been refuted by experiments on pigeons. Möbius urged that birds were guided in their journey by the direction of the roll of the waves ; Newton replies that though this may be a constant direction in certain parts of the Pacific, it is most inconstant in the stormy North Atlantic (37).

There is a very generally accepted idea that birds prefer to travel with the wind striking them diagonally — the " beam-wind theory," a theory, which so far as I can see, has absolutely no sound foundation. When on the Kentish Knock Lightship, Mr Eagle Clarke noticed, on a bad day, east to west migrants hurrying past " as if to avoid as much as possible the effects of the high-beam wind."

Mr A. H. Clark worked out the long oversea and overland course followed by the American golden plover, and showed to his own satisfaction that the birds always travelled at right angles to the prevailing winds ; therefore, he argued, they were guided by the beam-winds; always keeping the wind on their flanks led them aright (14). He says that if they fly at 100 miles per hour, with a beam-wind of 30 miles per hour, they will reach a spot 100 miles from whence they started, but 30 miles to leeward of a line drawn at right angles to the wind. Thus, if they rely upon the wind, their course is more or less diagonally across it according to its strength. He maps out the supposed route according to prevailing winds, but fails to notice that the very route he maps may be caused simply by the leeward drift when flying on winds which are not with them. One portion of the journey is enough to illustrate what I mean. From Labrador to the east of Bermuda the birds fly south-east, so, he argues, as to cross the south-west wind at right angles. But supposing the birds headed due south, meeting the south-west wind on their right front, they would of necessity, if the wind was strong, drift away to the east. It is improbable that they actually aim to strike the Bermudas, for it is only during certain weather conditions that they visit these islands. In favourable weather the birds do not touch the Bermudas, but continue their flight direct to South America.

The leeward drift of birds in a strong beam-wind may be noticed during ordinary flight; it has occasioned one of the most remarkable of Gätke's

statements. Referring to hooded crows, he says—
" To escape the disagreeable experience of having
the wind (south-east) blowing through their plumage
obliquely from behind, they turn their body southward, and appear to be flying in this direction.
This, however, is not the case. They do not make
the least forward progress to the south, but their
flight is continued in as exactly a westerly course,
and with the same speed, as though the birds were
moving under favourable conditions straight for-
wards, *i.e.*, in the direction of the long axis of the
bodies. This is shown in the most convincing
manner by such bands as happen to pass immediately over the head of the observer.

" Besides hooded crows, many other, indeed
perhaps all species, are capable of executing a later-
ally-directed movement of flight of this nature,
not only under such compulsory conditions as they
may encounter during the flight of migration, but
also during the ordinary activities of their daily
life " (**29**). He admits that he once thought it was
a drift to leeward, but that he is now convinced that
it is intentional, and is sure proof of his East to West
flight. In the face of such absurd statements as
these, how can anyone quote Gätke as an authority
on migration ! Yet, in recently-published books,
this east to west flight across Heligoland to Yorkshire is stated to be a proved fact, though Mr Eagle

Clarke, so long ago as 1896, showed it to be un-supported by British evidence.

Drs J. B. Watson and K. S. Lashley's experiments (57) of releasing marked birds, carried from their homes in closed cages, provides a strong argument in favour of orientation. They took fifteen sooty and noddy terns from Bird Key, Tortugas, and liberated them at intervals after they had been marked. The shortest distance was 20 miles from the Key, the farthest, Cape Hatteras, 850 miles ; thirteen returned to the Key. Neither sooty nor noddy terns range, as a rule, north of the Florida Keys, so that it is unlikely that any of the birds had been over the route before. They could have gained no experience, or hereditary knowledge, and as they were released during the breeding season, there would be no marked movement southward which they might follow, nor would they at that time be impelled by any desire to migrate. The change of direction from the Florida Keys, westward, to the Tortugas, occasioned by the water course which feeding habits would force them to follow, " removes the direction of the wind as a guiding agency, whilst the absence of landmarks over the great portion of the journey makes it improbable that sight was of service in finding the way."

CHAPTER VI

NOT only do the distances of the migration paths of different species vary considerably, from a trip of a few miles to a voyage from the Arctic to the Antarctic, but the individuals of one and the same species do not all travel to the same degree.

The familiar swallow, *Hirundo rustica*, though subject to certain geographical variations, is found throughout the Palæarctic and Nearctic regions, nesting throughout Europe to between 63° and 70° north and in Africa north of the Sahara, where, however, Canon Tristram found it also wintering in the oases. South of the Sahara to the Cape it is a winter visitor. In Asia it breeds, according to Seebohm, in Asia Minor, Persia, Afghanistan, and western Siberia, and winters in Scinde and western India. One form breeds in and north of the Himalayas, eastward to China and Japan, and winters in India and Burma, and another ranges from eastern Siberia across Behring Strait throughout North America, so far south as Mexico. This form winters in Burma, in Central America and

E   65

Brazil, but the Mexican birds are more or less stationary at all seasons.

Our swallow and its congeners have an almost cosmopolitan range, summering in the Northern and wintering in the Southern Hemisphere or comparatively near to the Equator in the Northern. Towards the centre of its range its migrations are either short or the bird is non-migratory.

Mr W. L. Sclater, addressing the South African Ornithologists' Union (**42**), stated that the swallow arrives at Cape Town at the end of October, and is common from November to March ; practically all have left by the middle of April. Swallows begin to arrive from the south in Africa north of the Sahara in the latter half of February ; early in March they reach southern Europe, later in the same month they are in Central Europe and by the middle of April large numbers arrive in England. Thus some swallows leave South Africa after others have arrived in England, and we might argue that this supports the thesis that those which go farthest south in winter breed the farthest north. The recovery in South Africa of some eight swallows ringed in the British Isles proves that some birds, at any rate, which do not nest in the most northerly parts of their range reach the farthest south. The finding of a swallow, that had been ringed near Cardiff, in Belgian Congo in December, where, as

Mr Witherby points out, it was evidently in winter quarters, is an interesting illustration of variability in habits. We cannot say by which route our birds reach South-East Africa, but it certainly suggests that all the birds from one area do not seek the same winter quarters. When the swallow reaches South Africa it is in ragged, worn plumage; before it begins its northward journey it has moulted.

Waders and shore birds which reach South Africa in autumn—the spring of the Cape—are moulting into winter dress; before they leave they have often assumed or partially assumed the breeding dress. When they arrive the native South African birds are breeding, but though Mr Sclater thinks that some nest a second time in the south, no satisfactory evidence has ever been brought forward to support the suggestion. These long-distance travellers not only move from a zone of moderate temperature to a warmer one, but many of them pass through the hotter zone to a country having a similar temperature to the one in which they bred, thus enjoying summer but not torrid heat all the year round.

There are birds in which the northern and southern forms are distinct. The wheatear, *Saxicola oenanthe oenanthe,* reaches us early, sometimes during the second week in March, and speedily settles down to nest. Towards the middle or end of April a brighter larger bird appears, the Greenland wheatear, *Saxicola oenanthe leucorrhoa*, which was recognised in Greenland, Iceland and eastern North America before it was seen that both forms occurred in Britain. This larger bird often loiters through Britain, for its northern home is not ready for it until the Arctic spring. We know it breeds farther north than our wheatear, but its winter range is not fully worked out. The larger bird is known in winter in North-Western Africa ; the winter range of the smaller form extends eastward and southward to the Zambesi. Both occur in Senegambia. Thus the northern bird, so far as we know, does not winter so far south as the other.

The folly of laying down the law on the strength of the knowledge of the habits of a few species is shown by the study of the movements of American birds. Mr Cooke shows that as a rule " the migration is a synchronous southward movement of the whole species " in autumn, " the different groups of individuals or colonies retaining in general their relative position." The black and white creeper *Mnistitta varia* breeds from South Carolina to New Brunswick, nesting in the south in April and reaching the northern limits in the middle of May. In the middle of July old and young birds have been seen at Key West, 500 miles south of the breeding range, and towards the end of August they have reached the north coast of South America. The New Brunswick birds cannot be ready to leave before the middle of July, and Mr Cooke allows them fifty days for the trip, bringing them to the Gulf States in September ; he argues that this is proof that the earlier migrants must have been birds from the southern part of the range. Black-throated blue warblers, *Dendroica coerulescens,* reach Cuba at about the time that others of the same species are arriving in North Carolina ; the first, he concludes, are birds from the southern Alleghanies and the others from northern New England or beyond (20). Other species illustrate the same order which he calls " normal," but show that it is not an invariable rule.

Southern-bred Maryland yellow-throats, *Geothlypis trichas,* reside throughout the year in Florida ; those in the middle districts of the range migrate for a short distance only, whilst the Newfoundland birds pass over the winter home of their southern relatives to the West Indies. The palm-warblers of the interior of Canada travel 3000 miles to Cuba, passing through the Gulf States early in October ; those from north-eastern Canada travel later and slowly and settle in the Gulf States, after a journey of only half the distance. He sums up wisely— " No invariable rule, law, or custom exists in regard to the direction or distance of migration. . . .

Each species presents a separate problem, to be solved for the most part only by patient, painstaking observation and by the recognition of subspecies."

The order in spring is yet unproved. "With many birds . . . the first individuals to appear in spring at a given locality are supposed to be old birds that nested there the previous year." These are followed by those which nested a little farther north, followed later by those whose homes are in the most northerly part of the range. "If, then, for any species, the southern nesting birds lead the van in both fall and spring migration, and the near guard in each case is composed of northern breeding birds, it follows that some time between October and April a transposal of their relative positions occurs ; and that the more southern birds pass over the more northern ones, which delay their migration, knowing that winter still holds sway in their summer dominions." It is not known where this transposal takes place, nor whether the northern birds remain in winter quarters till the southern birds have passed, or start a slow migration, during which the southern birds pass over them. Later another transposal occurs ; the northern birds cross the southern part of the range, passing birds which are already nesting. "Spring migration seems to be therefore for some species a game of leapfrog—the southern birds first passing the northern, and the northern passing them in turn " (20).

The custom, now fortunately becoming widespread, of marking birds by affixing a numbered metal ring to one leg, may help to elucidate this and many other problems, but until a large number of results are collected it is unwise to draw conclusions. Almost every month the recovery of some of these marked birds is noted in the scientific journals, but so far, beyond indicating the minimum distance travelled by individuals, little can be proved. It is, however, plain that birds do not invariably act as they ought to do if they obeyed all the laws which have been invented for them. A few records or results may be quoted, but any suggestions from these must be treated as suggestions only ; many more must be forthcoming before we can say, proved.

For about thirty years the white stork, *Ciconia alba*, has been systematically ringed in Denmark, Germany and Hungary, and Dr A. L. Thomson (58) gives a valuable summary of the results obtained by Mortensen, Thienemann, Schenk and others. Herr Schenk showed me in Budapest his elaborate methods of registration, and I accompanied him on ringing expeditions, but I have not got his most recent recoveries. Young storks, taken during their first autumnal journey, show a general south-

easterly trend through Europe. Danish birds have
been recovered in Germany, Poland and Hungary ;
German birds in South-East Germany, Czecho-
slovakia and Hungary ; and Hungarian birds in
the Balkan States. On return migration there are
records from Syria, Palestine and Asia Minor.
Storks marked in Germany, notably in the south-
west, have been recovered in Spain and southern
France, even far west, and a Danish bird was found
in spring in Algeria. Dr Thomson concludes that
results indicate a south-eastern migration from
Northern and Central Europe, and a south-westerly
trend from the more westerly parts of Central
Europe. The Danish bird in Algeria, and a bird
recovered at Barcelona which had been ringed at
Cassel, not far from the starting-point of birds
which took the south-east route, are puzzling, but
further records may explain these exceptions.

The long-distance records are, with the exception
of an Hungarian bird recovered in South-East
Arabia, all from Africa, and become most numerous
in the south-east, until in the eastern part of South
Africa there were, when Dr Thomson collected his
data, sixty-five recoveries so near together —
twelve Danish, twenty-one German, and thirty-two
Hungarian birds—that they could not be shown
individually on a small scale map.

Dr Thomson made special study of the movements

of other species, collating all known records ; these are the swallow, lapwing, starling, mallard, pintail, herring gull and lesser black-backed gull. In another chapter he gives the general results of bird marking in Europe and America. From this and later reports I have extracted a few long-distance records. Common terns ringed in Germany and Sweden have reached Natal and the Cape, one ringed in the North-Eastern States was found in Trinidad, and another, ' banded ' in Maine, three years after ringing, at the mouth of the Niger. German, Dutch and British Sandwich terns are recorded from West Africa, the Cape and Natal, and black-headed gulls, marked in Germany, have crossed to Barbados and the eastern shores of Mexico. A British ringed bird was found in the Azores, where, also, a Dutch spoonbill was recovered. Two kittiwakes, marked as nestlings on the Farnes, were recovered respectively in Newfoundland and Labrador, both in their second autumns.

Much has been learnt from the intensive study of one species, but the results apply to that one only ; each species must be treated separately. In time this will be done. In the opening chapter I mentioned the uncertainty about the behaviour of any individual song thrush, merely as an example. So far, the few records of marked song thrushes add to rather than solve this problem. Years ago the song thrush was looked upon as a permanent resident so far as Britain was concerned. Now we know it to be either migratory or sedentary even in Britain. What do we find ? A song thrush, marked as a nestling in July in Northumberland, is found in November in Durham ; another one, marked in Berkshire, travels to Norwich and is recorded in November ; but a third, born in Aberdeen, takes an autumnal flight of at least 1500 miles and is found in Portugal. Several British thrushes have been found in western France and the Peninsula.

It is said that home-bred lapwings are somewhat sedentary, and that the large winter flocks are composed of Continental immigrants. But lapwings marked as nestlings in the British Isles have been recovered in the south of France, Spain, Portugal and Morocco, and many in Ireland. Dutch birds have travelled to France and the Peninsula, and one to Morocco ; Hungarian birds have been found in Italy, France, Spain and Algeria. Swedish, Danish, Dutch, German and Esthonian lapwings have been recovered in England. This is true, so far as it goes, but a lapwing marked as a nestling near Stirling has been found in the south of France, and two others in Portugal, whilst five have been recovered in Ireland.

The results of marking seabirds are interesting, showing that the young birds often wander northward in search of food before there is any marked autumnal southward migration. Terns and black-headed gulls have been found a month or more after they have left the nest to the north of their breeding colonies in Cumberland and mid-Wales. A bird from Ravenglass was taken in its first January in Brittany. Rossitten black-heads have been shot in the Isle of Wight and in Breydon in Norfolk.

This may only mean that the young blackhead is a confirmed wanderer in search of food, but the few results with woodcocks, marked as British-bred nestlings, are puzzling. They have been known to linger in the neighbourhood of their home until November, and have been found in Portugal only a month later. Birds marked at Tyrone have been found so far apart as Cornwall, Harrow and Inverness; what route for the Irish birds can be guessed at ?

Birds marked as adults present further problems, but also provide interesting evidence. Hooded crows, captured on migration in spring at Rossitten and then released, have been recovered in autumn actually in the same place and in other localities in Germany, and one marked in October was taken two years later, in spring, in Finland. The sum of these records of crows proves one thing con-clusively—the fallacy of Gätke's due east to west

and west to east flight, and supports a coastwise
migration for this species.

Adult teal, captured in decoys, ringed and released in South Denmark in September and October,
were taken in November and December in Hamp-
shire, Suffolk and the Moray Firth, whilst others
from the same place were recorded from other parts
of England and Ireland, from western France,
Holland, the south of Spain and the north of Italy.
Fly-lines, if followed, are divergent and complicated.
Four young herons were marked in one nest in
Denmark ; one was recorded in Holstein in June,
and another in Mecklenburg in July ; the third
was killed near Salisbury in Wiltshire in October,
and in the following February the last was obtained
in the north-west of France. Two from another
nest were recovered in Denmark, one in July and
the other in February, twelve months after birth.
Another heron reached Andalusia by August. In
each case where there was indication of a direction
it was south-westerly. Many more records might
be mentioned, but these are sufficient to show the
value of the method and the present insufficiency of
results.

Many of these records show that the speed of
the migrating birds, even in spring, is not great.
Mr Cooke proves that most species in North America
travel slowly through the districts where food is

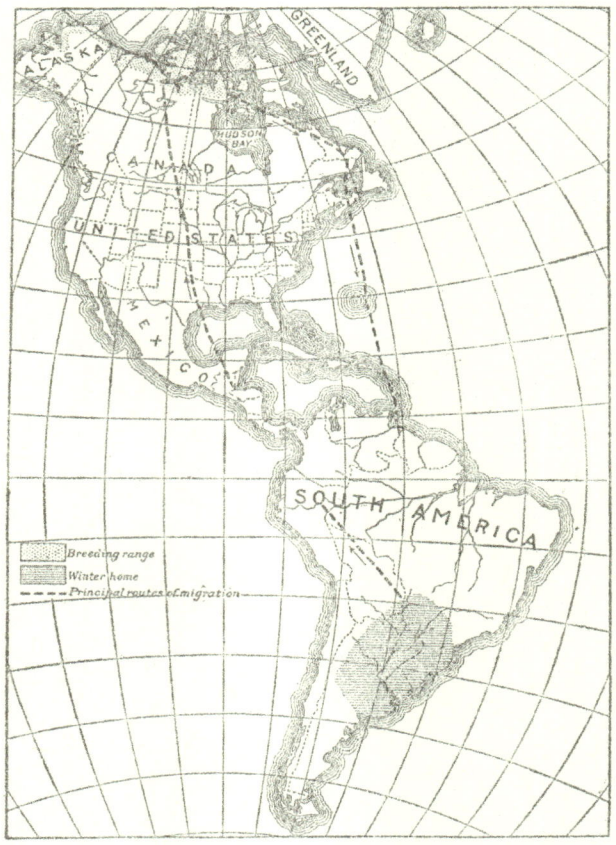

Map showing the range of the American Golden Plover,
with its known migration route.

(From *The National Geographic Magazine.*)

plentiful and during the earlier part of the journey northwards only a few miles are covered per day ; they travel with the slowly advancing vernal wave, but, as we shall show in the next chapter, many species actually outstrip it, and travel from warmer to colder climates.

By the kind permission of Mr Cooke I am able to reproduce three of his maps, illustrating the longest known distance travelled by any bird in a single flight, and the probable evolution of this extraordinary oversea voyage (**21**). This long journey, roughly 2500 miles at a flight, is used in autumn by several species of American shore birds, and the particular species most easily recognised, is the American golden plover, *Charadrius dominicus*, which differs but little from our *C. pluvialis*. An important point to notice is that the route followed in the fall is not the one used by the bird in spring, an undoubted proof that all routes are not identical with the original line of dispersal of the species. Nor is the route directly from the north to the south, though there is plenty of evidence to show the fallacy of the notion that all birds move in this one direction.

The golden plover nests along the Arctic coasts of North America from Alaska to Hudson Bay. So soon as the young are able to take care of themselves the birds migrate south-east to Labrador,

where for some weeks they fatten on the autumn
harvest of fruits. A short journey across the Gulf
of St Lawrence brings them to Nova Scotia, where
they gather before starting on their oversea flight.
The eastward trip to the food-supplying districts
is support of the idea that a route is originated by
passage from food-base to food-base, rather than
by any hasty rush from the dangers of approaching
winter. The birds start south from Nova Scotia
for South America !

During this long oversea journey, which Mr
G. H. Mackay thinks, with reason, may be under-
taken under favourable conditions at a speed of
from 150 to 200 miles an hour by birds with such
magnificent power of flight, the plovers may meet
with many different winds. The Cape Cod sports-
men look for them if the wind is strong from the
north-east ;   the Barbados gunners expect them
when there is squally weather from the south-east,
but when westerly breezes are blowing they will
pass so far as 400 miles east of the Bermudas.
Only when the wind is adverse and strong do the
plovers visit the Bermudas or even stop at any of
the northern Lesser Antilles, 600 miles from the
coast of South America. In favourable weather
they neglect any of these " emergency stop-overs "
and hasten on. In the Guianas the birds rest and
feed, but they soon move on. Across the Brazils

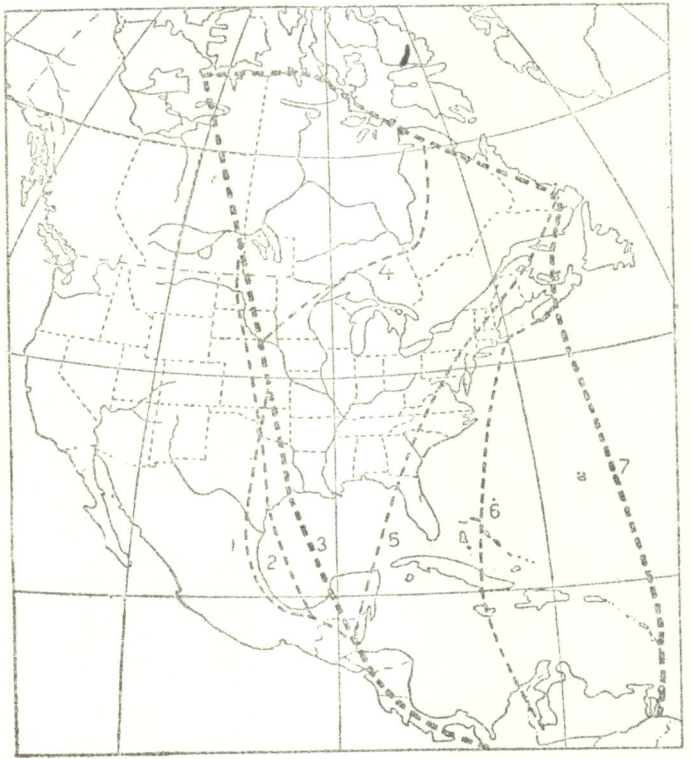

Map showing the evolution of the migration route of
the American Golden Plover.

(From *The National Geographic Magazine.*)

their actual route is uncertain, but they have been met with in Amazonia, and are known to winter in Argentina, and, it is suspected, in eastern Patagonia.

The return migration is, so far as it is known, in a steady northerly direction, rather north-west across Bolivia towards Central America. From Yucatan they cross the Gulf to Texas, then slowly travel up the great Mississippi highway and across Canada to their northern breeding grounds. " Its round trip has taken the form of an enormous ellipse with a minor axis of 2000 miles and a major axis stretching 8000 miles from Arctic America to Argentina."

The following is Mr Cooke's suggestion of the origin of this great ellipse. Towards the close of the glacial era, when the ice began to recede, the Florida peninsula was submerged and only a small area in the south-east of the States was free from ice. Plover attempting to follow up the retreating ice were confined to an all-land route from Central America through Mexico to the western part of the Mississippi Valley. As the east gradually became uncovered the route would be extended to the north-east, until the area stretching to the Great Lakes was fit for bird-habitation. As the route lengthened and the power of flight developed, there would be a tendency to shorten the line by cutting off some of the great curve (No. 1) through Mexico and Texas, and a short flight across the Gulf (No. 2) would be gradually lengthened, until the present spring route, then also the autumn route (No. 3), was attained. As Canada opened out, the routes in spring and autumn diverged; in autumn the fruits of Labrador were an attraction, but the Chinook winds made the country east of the Rockies more suitable for spring migration ; the fall route tended eastward (No. 4), the spring route remained unchanged. When the fall route had worked eastward to the Gulf of St Lawrence (No. 5), shortening took place in the same way from the great westward curve, culminating in an ocean flight, short at first (No. 6) and later extended, the total distance shortened, until the present route was attained (No. 7).

This reasoning, sound enough, helps to a more difficult problem—how the Pacific golden plover, *Charadrius fulvus*, found its way to the Hawaiian Islands, where numbers of the birds winter annually. Roughly the islands are 2000 miles from California, 2400 from Alaska, whence the birds fly, and 3700 miles from Japan. Mr Cooke scouts the idea that any bird flies aimlessly out to sea to find a new winter home, and the chance colonisation by a storm-swept party is as improbable ; if this did occur it is hardly likely that they would at once

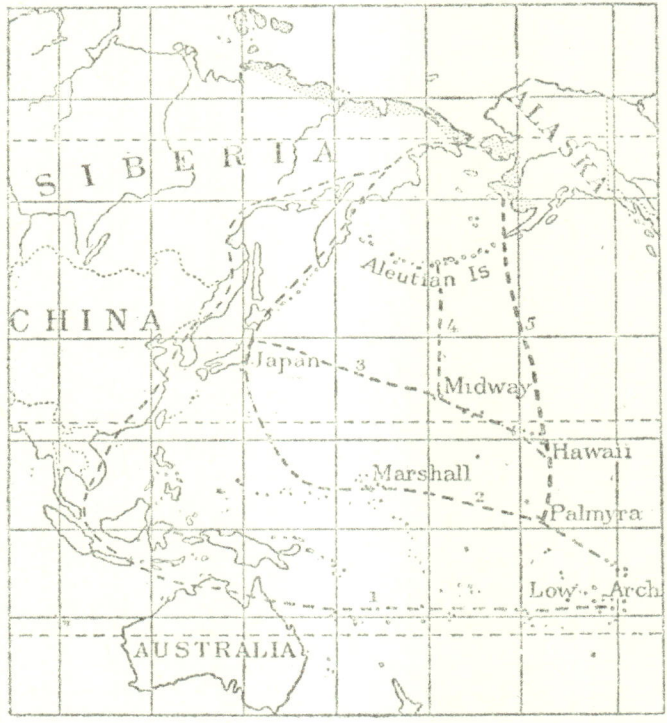

Map showing the evolution of the migration route of
the Eastern or Pacific Golden Plover.

(From *The National Geographic Magazine.*)

depart, in a single season, from ancient habits and carve out an entirely new migration route. Probably the origin of the route is as follows. The bird breeds on the northern shores of eastern Siberia from the Liakof Islands to Behring Strait, and on the Alaskan side south to the northern base of the Alaska peninsula. It winters on the mainland of south-eastern Asia, in eastern Australia, and throughout the Oceanic Islands from Formosa and the Liu Kiu Islands on the north-west to the Low Archipelago in the south-east.

It is fairly certain that the original route would be roughly north and south, between Siberia and southern Asia. In time the species spread eastward in winter, to Australia and to islands farther east, whilst the breeding area extended to Alaska. If these extensions took place before any cutting off of corners in the route, Alaska birds would travel 11,000 miles to reach the Low Archipelago, only 5000 miles in a direct air-route (No. 1). Probably shortening began early among the Pacific islands, from the northern islands to the Asiatic coast, and finally to Japan (No. 2). From Palmyra the flight to the nearest of the Marshall Islands is 2000 miles ; thence a journey, provided with several possible rests, of 3000 miles would bring them to Japan. A thousand-mile drift through strong winds might cause the birds to reach Hawaii, whence they would

F

find a chain of islands which would help them, and render the last flight to Japan no longer than the one they had been accustomed to. Having once reached the Midway Islands the shortening of the route would be carried on again by lengthening the oversea journey northwards until the Aleutian Islands were discovered (No. 4). The present route, now followed in spring and autumn (No. 5), would be the natural climax of this long evolution. The two golden plovers, sub-specifically distinct, nest little more than a hundred miles apart ; their migrations and winter homes are as different as they could be in any two widely divergent species. It is one of the most striking of the ascertained facts in the distribution and habits of birds.

Two long-distance flights, mostly overland, are those of a wigeon and a redwing. The wigeon was caught as a mature bird in autumn, and was recovered near Uralsk, north of the Caspian Sea, some 2000 miles east, over two years later. The redwing was ringed in Finnish Lapland and was found in Italy.

CHAPTER VII

MIGRATION AND WEATHER

IN previous chapters it has been necessary to refer repeatedly to the connection between migration and meteorology ; either the relation of periodic movements to the rotation of seasons, or the influence directly or indirectly of weather conditions upon normal and abnormal migration. That there is an overruling relation between the advance of spring and the passage to northern breeding quarters, and the gradual cooling in autumn and the retreat to winter quarters is, of course, evident, but it must not be held, as contended by the early students of migration, that this is the sole factor which regulates migration. The actual relationship between the weather and the movement of birds is far more complicated than one would imagine, and the stimuli of continental or overland travelling differ from those of a cross-sea flight.

In the British Islands most of our larger movements are at their start or their finish, or both (so far as our area is concerned), oversea passages, and unless the weather be absolutely favourable, birds do not undertake these voyages. No one has added more to our knowledge of the connection, in what we may term British migration, than Mr Eagle Clarke, but it must not for a moment be imagined that his conclusions and the data from which he arrived at them are purely insular. The British Islands are merely the field of observation, the centre of the field, of the movements of Holarctic birds which travel regularly or occasionally through Britain. Mr Clarke points out repeatedly that in studying the phenomena it is the conditions at the point of departure not at the point of arrival—generally the point of observation—which are important.

The oft-repeated assertion that birds can foretell the nature of approaching weather—that they are living barometers—is not supported by any satisfactory evidence, but it is certain that on many occasions the weather into which they have passed in moving from one zone to another has not only retarded, checked, or exhausted them, but has proved fatally disastrous. During the westward rushes in winter, when exceptionally severe weather has cut off the food-supply of ground-feeding birds, observers who have seen the birds moving in front of the storm have maintained that they had felt its approach and retreated in time. The truth seems to be that the birds start so soon as the supply is cut off but in many cases speedily outstrip the storm. When these exceptional winter migrations take place the birds in the lowlands of Lancashire and Cheshire move westward towards Ireland, and are observed at different points along the North Wales coast. They are sometimes seen travelling in a snow-storm and sometimes in advance of it. In eastern Cheshire I have seen parties of lapwings passing over westward just in advance of snow, which when it reached the East Cheshire fields, started the local lapwings after their relatives from farther east.

During regular migration birds start in favourable weather but frequently meet with unfavourable weather before their arrival at the point aimed at ; most of the bird " disasters " at the lighthouses and lightships, and more occasionally inland, can be explained in this way.

In his digest of the observations at lighthouses and lightships Mr Eagle Clarke shows that spells of genial weather are favourable and that during these spells migration is even flowing and continuous (15). Slightly unsettled conditions have little effect, but an increase of the irregularities accelerates migration. Sooner or later cyclonic disturbances interrupt regular movements, and, if these are extraordinary, act as barriers, either holding the birds in one place or forcing a hurried departure or " rush." Favourable weather immediately following a check or " hold up " often causes a rush ; a sudden fall in temperature may force large numbers of birds on in autumn or retard them in spring. Temperature, he declares, is the main controlling factor in all extraordinary movements, other meteorological conditions being suitable.

In the autumn migration to Britain, the chief movements take place when a large and well-defined anticyclone has its centre somewhere over Scandinavia, with gentle gradients in a south-westerly direction over the North Sea. Coincident with this we usually find cyclonic conditions prevailing to the west of the British area, with low-pressure centres off the west or south-west of Ireland. The weather is clear and cold, with light variable airs over Scandinavia, but in Britain the sky is overcast, and the wind easterly and moderate to strong ; not infrequently these conditions mean fog on our eastern coasts. If the birds leave Scandinavia under favourable conditions they may be met by the approaching cyclonic system, which usually, though by no means always, travels in a north-easterly direction across the Atlantic. Migration is thus checked, but a return of favourable anticyclonic conditions starts the birds again, often with a fresh impulse in the shape of falling temperature. When the anticyclonic area is exceptionally large, extending from the Scandinavian peninsula in a south-westerly direction and embracing the whole of the British Islands, simultaneous immigration and emigration may be witnessed.

Cyclonic spells are not always unfavourable to migration. In spring, when they are of a mild type with soft rain and warm winds following after a cold anticyclonic period, a northward movement is frequent.

Mr Eagle Clarke says that the importance of winds is overstated, but as an incentive only. The direction of the wind has no influence as an incentive but its force is an important factor; in a strong wind a bird may be blown out of its course. Birds will not start in a high wind but may pass into the influence of strong winds which may affect both progress and direction. He adds that particular winds usually prevail during the season of great autumn movements, which are not incentives but are the result of pressure distribution which is favourable to migration. These are usually northeast to south, but a westerly wind would serve as well, but it indicates a pressure distribution which is fatal to migration between north-west Europe and Britain—cyclonic areas to the north-east and east of our area.

All this, no doubt, is perfectly true. It is founded on the analysis of a huge number of carefully recorded observations, and upon a general knowledge of migration which few can ever hope to equal. Mr Clarke understands his subject. It appears, however, to me that he may put rather too much weight upon the barometric influence, and too little on one side of the wind question. Are we yet in a position to say that birds do not make direct use of certain winds ? It may be that the use of the prevailing winds at migration time is far more unconsciously intentional (if such an expression can be used) than is at first apparent.

One or two points must be kept well to the front which are often ignored by observers. Firstly, very much visible migration is abnormal ; that is to say, most of the incidents of passage which are noticeable, especially observations at the lightships and lighthouses, are during spells of weather which are described as unfavourable ; it is the " hold-ups," checks, and " rushes," which attract attention far more than the even-flowing normal migration.

Mr J. Tomison, in his valuable notes on observations made at Skerryvore (52), shows that in ordinary clear weather birds pass at a great height, beyond the power of vision. He proves this by instances of the diurnal passage of redwings, birds which are generally supposed to migrate at night, and undoubtedly do so frequently. He heard the well-known passage-note in the daytime, but with the naked eye could see no redwings ; he found them with the telescope and later discovered others which were passing above the range of normal vision. Mr Eagle Clarke, commenting upon the extraordinary numbers of rare and exceptional visitors which are noticed on many islands—Fair Island, the Flannens, the Isle of May, and Heligoland may be taken as a few examples—says that it is their detached position and comparatively small size which makes these islands so useful to the observer. The same variety of birds and greater numbers reach larger islands and tracts of land, but they are unobserved when they are thinly distributed and not massed or confined in a small area. "With all our great army of trained observers," he declares, "we in Britain see only an infinitesimal number of the migrants which visit our shores . . . " and "this is especially the case on the mainland."

During an anticyclone there is a descending movement of air currents from a centre of high pressure in all directions, and these currents or winds are deflected "clockwise" in the northern hemisphere ; and when cyclonic conditions prevail the air currents are directed inwards towards a low-pressure central area, rotating spirally at the surface of the earth in the direction contrary to the hands of a watch. In the southern hemisphere the directions are reversed. A cyclonic system is usually carried forward by great drift winds like eddies upon a swift stream," in the North Atlantic as a rule from south-west to north-east.

Do we really know the force and direction of the winds at a high altitude during these movements ? Are we not merely guessing at the real aerial conditions by the movements near the earth at the time of the departure of the birds ? Is it fair, if I am right that the meteorological observations are founded upon only those observed at comparatively low altitudes, to lay down laws as to the particular conditions which are favourable or unfavourable, or the particular winds which are used or avoided ? The direction of the wind may be the same up to a great height, many thousand feet, or it may vary within 500 feet of the earth. Nearly fifty years ago, when Glaisher made his great ascents, he sometimes met with three or four currents moving in opposite directions. The more recent upper air investigations show that though as a rule the wind at various elevations is in the main from one point of the compass, its degrees vary considerably, and its force at the various heights shows remarkable differences. Generally the force rises to about 5000 feet, but there is no invariable rule. I tabulate a few examples taken more or less at random from the Weather Reports for 1908. The altitudes above the ground are measured in metres, roughly converted into feet ;

the letters indicate the direction of the wind, the figures its speed in miles per hour. The last one in the table, observations made at Brighton on September 20th, is particularly useful. The conditions on this date were anticyclonic, and favourable to migration. At 400 feet above the sea the wind was blowing at 5 miles an hour ; at between 5000 and 6000 feet its force was 20 miles per hour. What then would happen to a bird leaving Brighton for say the Spanish Peninsula ? If it flew at 20 miles an hour towards the French coast about Dieppe, it would meet the wind blowing at 5 miles an hour, and take between five and six hours to reach the coast, head to wind. If it rose to the height of 3000 feet it would meet a wind blowing at the same speed as it was flying, and it could make no headway. If, however, it flew in a south-westerly direction the more it turned westerly the farther it would drift down channel towards Normandy or Brittany, and be carried out to sea ! But this is exactly what would not have happened, for on this date a feeble cyclonic system was approaching from the Atlantic and extending its area of influence over southern England. In the Channel the bird would meet westerly winds which would bring it safely to the Brittany shores, or if it missed them, to the western shores of the Bay, where the wind was actually from the north. I mention this

Date.	Station.	Ground Level.	100 mtrs. (330 ft.).	500 mtrs. (1660 ft.).	1000 mtrs. (3320 ft.).	1050 mtrs. (5000 ft).
Jan. 2	Petersfield	NE by E	...	ENE 30	E by N 50	E½N 13
,, 2	Glossop 1100 ft.	E by N 8	...	E 15	E by S 30	
,, 3	Pyrton Hill 500 ft.	ENE 14	...	E by N 35	E 53	
,, 4	,,	NE by E 10	...	ESE 25	E by S 25	E by S 30
,, 11	Petersfield	S by E	...	S 10	SW by W 3	SW by W 5
April 9	,,	SE	...	N by W 7	...	N½W 20
,, 8	Glossop	N	...	N by W 9	NW by N 16	W 7
,, 30	,,	S 14	S by E 27	S 30	W by N 46	
May 16	,,	WSW 16	W by S 26	W by S 27	W 29	W by N 33
Sept. 5	,,	WSW 12	W by S 15	W by S 17	W by N 21	W by N 23
,, 7	,,	S by W 9	SSW 16	SSW 20	SW 33	
,, 10	,,	NW by N 8	NW by N 16	NW by N 21	NW 34	NW 36
,, 20	Brighton 380 ft.	ESE 5	SSE 5	S 15	SSE 20	SSE 20

2000 mtrs. (6660 ft.).	2500 mtrs. (8320 ft.).	3000 mtrs. (10,000 ft.).	3500 mtrs. (11,660 ft.).	4000 mtrs. (13,320 ft.).	4500 mtrs. (15,000 ft.)	5000 mtrs. (16,700 ft.).	6000 mtrs. (20,000 ft.).
ENE 23	NE 22	NE by N 18	NE by N 25	NE by N 23			
ESE 35	SE by E 20	SE by E 15					
N by W 9	NW 8	NNW 7	N½W 11	E½N 8	E by S 14	E by NE 13	ENE 14
N 14	NW½N 9	NW by W 12	...	NW by W 18	W by N 20		
NE 6	N by E 8	W by N 8	NNE 9	NW 1	SSW 3	NNW 5	SW 7
WNW 36							
W by N 28							

merely to show that apparently unfavourable winds may be really favourable.

Under ordinary circumstances are we justified in saying that birds make use of the winds blowing with a certain force at the point of departure, or that they ignore them ?   Certainly we cannot judge by either the force or direction of the wind at the point of arrival, as Mr Clarke points out.   The bird may have dropped into most adverse currents.

In Hungary, where migration has been very carefully studied, we find evidence supporting Mr Clarke's theory, and yet giving it a slightly different complexion. Low atmospheric pressure, depression (the warm cyclonic conditions of spring) very often shows the greatest rate in the arrival of the swallow. If there is a centre of depression west of Hungary, and its path is directed north or north-east, swallows appear in crowds.   The fair side of the depression, with its warm southerly winds, is therefore favourable. A list of twelve other birds, which also appear in spring under these conditions in greatest numbers, is added. The " bad " side, with cool northerly winds causes delays in the arrival of these thirteen species. The depressions often have a sphere of influence extending so far as North Africa, so that birds, on the fair side, can cross the Mediterranean with southerly winds all the way (31).

I have endeavoured to show that often the force
of wind is greater at a high than a low altitude, and
there is ample evidence to prove that birds fly
at a great height when conditions are favourable.
Birds usually leave Scandinavia when there are
descending currents flowing outwards from the
centre of high pressure ; is it wild speculation to
suggest that it is the southward flowing currents,
which are also deflected westwards, upon which
the birds intend to travel ? Thus the bulk of the
Scandinavian birds might not touch Britain at all,
but those which started upon light to moderate
north-east to easterly winds from the western
shores of Norway would be helped to Britain.
Mr Clarke mentions that when he was at Fair
Island, north-west to westerly winds did not stop
migration from the north, but is it certain that
the birds did travel in or against these westerly
winds ? May they not actually have travelled on the
" good side " of the cyclonic system, with these very
winds carrying them towards Fair Island ? their
actual visible approach from the north does not prove
that they had travelled all the way in this line.

On September 22nd, he says—" The favourable
meteorological conditions of yesterday—fine weather
and moderate south-east breezes,—has had a marked
effect, for to-day goldcrests are swarming every-
where." But what does he mean ? Favourable

to him as an observer or to the goldcrests ? Surely the birds did not aim for Fair Island ; were not these weak-winged birds probably making for the south, when the south-east wind caught them and drifted them to the west ? Fair Island was a refuge, but hardly the objective of their flight (**17**).

Compare this with Cordeaux's notes of another goldcrest immigration, this time to the Lincolnshire coast (**23**). On October 13th the wind was north to north-east in the afternoon, light but increasing in force, the weather clear and bright—a few birds arrived. They had started under favourable circumstances. Shortly after midnight on the morning of the 14th, the wind got full east, with quite half a gale and heavy beating rain, continuous to the morning of the 16th ; the nights were very dark. " During this time the immigration was immense," and most of the birds were goldcrests. Cordeaux's idea that these were not normal immigrants but birds which were passing probably from north-east to south-west, when the easterly gale caught them, is probably correct.

I have referred to birds starting at a high elevation. Service says that in normal departure from the Solway, most birds mount to a high altitude, but " a strong beam wind will bring the birds— even those of strongest power—down to 200 to 500 feet of the surface, and it is interesting to see whole flocks with heads turned almost completely to wind, and yet travelling along at nearly their normal speed, at right angles to their position " (46). Mr Tomison mentions rooks, daws and hooded crows driven to Sule Skerry by south-east winds in March, leaving two days later in a westerly gale. They, at any rate, did not object to a strong wind which was in the right direction.

I have mentioned Mr F. J. Stubbs' paper on the " Use of Wind " (50), and I believe that there is much more in it than is actually proved by low-level observations. I doubt if birds always intentionally make use of strong winds, currents which would carry them for great distances at a considerable speed, but the preliminary ascent may be to search for these currents. Cyclonic and anticyclonic winds, even when at an altitude of some thousands of feet, would carry them easily, and probably it is the wind-borne individuals, parties, or even hosts, which drop for a refuge to the first island they see when carried far from their migratory path. They are carried rather than drifted from their pathway, borne in the moving current whether they wish it or not. Provided that the cyclonic winds are fairly steady in direction and force, sweeping round and inwards towards their centre, we may in imagination trace the pathway of our so-called lost wanderers to far distant islands ; without many more upper-air observation stations, we cannot actually prove the route.

G

But even putting aside the high altitude idea, and confining our route-tracing to the known courses of air currents, we shall find immense difficulty in mapping out the actual course of any bird on any particular day. The study of some of the publications of the Meteorological Committee, such, for instance, as the " Life History of Surface Air Currents," by Shaw and Lempfert, published in 1906, shows the great variation in the pathways, speeds, and formation of these systems ; a bird which accidentally entered a cyclone would unconsciously alter its actual track and speed very many times before it passed beyond the area of influence.

I am indebted to Mr Stubbs and Mr Herbert Taylor of King's College, London, for some interesting mathematically worked-out routes of birds, travelling at a given speed in a cyclone rotating at given speeds and moving at a fixed rate ; these show great variation both in direction and speed according to the time and place of entering the system. The track of the bird is, of course, influenced by its own rate of progress, by the speed of the rotating currents, and by the rate at which the whole system moves in any direction. Thus a migrant passing south and coming within the influence of a cyclone which is moving north-east at a high rate of speed, say 40 miles per hour, will, if it enters towards the northern limits of the system, be at first retarded by the con-

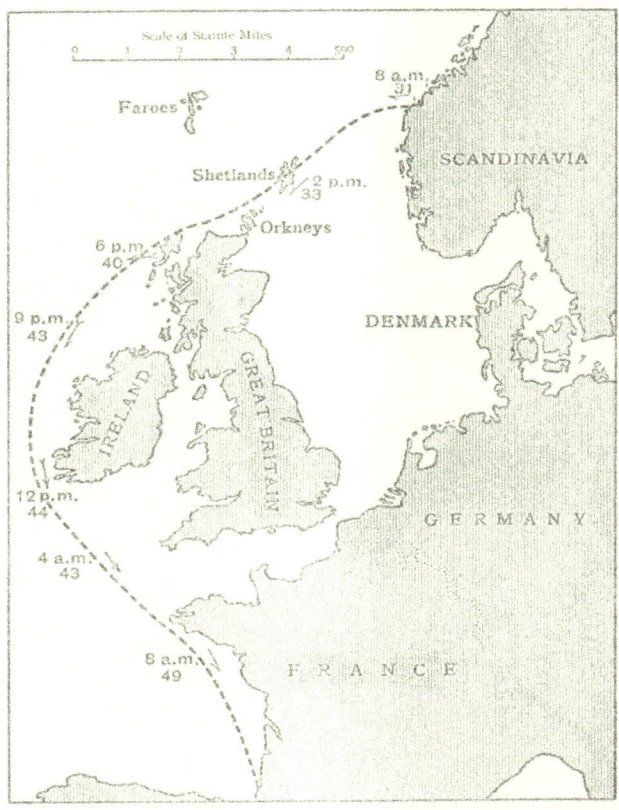

Map to show that a bird leaving Norway, near Aalsund, might be carried round the British Islands in twenty-four hours. The arrows indicate the actual directions and force of wind at the times marked during a slow-travelling circular storm in autumn 1901. Speed of bird about twenty-five miles per hour.

flicting forces of the easterly winds, the trend towards
the north-east of the rapidly travelling cyclone and
its own southward flight.   If it is flying faster than
the speed of the cyclone it will drift westward but
gradually approach the low pressure centre.   After
passing this its course will at once change and its
speed will be accelerated towards the east.

Even violent storms move at varying rates, and
it is conceivable that a bird leaving Scandinavia on
favourable anticyclonic winds might at once come into
the influence of a large, slowly-moving, circular storm,
with a low-pressure centre to the west of Ireland,
and might, if the air currents were strong, be carried
westward at first, then south and finally eastward, so
that it would actually pass round the British Islands.
I have taken this exceptional case from the actual
course of a storm, which varied between forces 9 and
11 on the Beaufort Scale (say an average of 50
miles per hour) but only travelled slowly eastward
at about 17 miles per hour.   In some cases the
storm centres are nearly stationary for many
hours.

It is easy to appreciate Herr Herman's statement
that spring immigration in Hungary is accelerated
on the good side of a mild cyclone ; the direction of
the bird, of the circulating air currents and of the
whole system may be coincident.   Given a low-
pressure centre west of the Bay of Biscay, spring

migration would be accelerated through Spain and France towards Britain.

Mr Stubbs points out that the pathways of several birds, or parties of birds, which started at different hours, would be divergent, for they would come within the influence of winds blowing in various directions according to the position of the system ; this he argues is contrary to the accepted idea of routes. This, however, entirely depends upon what we mean by a route, as I endeavoured to show in an earlier chapter. The journey from point to point is a route, although the bird may be drifted many miles in one direction or another on the way ; it is only when the bird fails to reach its objective, a suitable breeding place or winter station, that the route is a failure.

The frequent occurrence of rare birds, some of them almost or quite unknown elsewhere in Britain, on out-of-the-way islands, has led to strange theories. One is that there are regular fly-lines over Fair Island, the Flannens, St Kilda and elsewhere, similar to the one which is said to pass over Heligoland. Dr Eagle Clarke, whose work on Fair Island, St Kilda, and the Flannans has brought so much light, can give no other explanation than that these stragglers have failed to use the faculty of orientation, and are lost. The suggestion I can offer at present is that there are ornithologists directing their attention to these spots which, through geographical position and isolation, are the likely refuges for wind-borne migrants. Also that the accidental departure from the directions aimed at by the birds is, where wind and barometric systems are so variable, far more frequent than is usually suspected. Direct routes are doubtless aimed at, but only accomplished under favourable conditions for the whole journey. Migration is not infallible ; it is an evolving habit, strengthened by those which survive its perils, now as it was in its early days.

During a long overland journey, winds will probably have less influence, though for rapid passages high flights certainly appear to be not uncommon. There is, however, another aspect of the connection between migration and weather which we have hardly touched, migration synchronal to the change of season. Mr Cooke shows that in North America the push forward in spring is not in most species so soon as the weather permits ; they do not actually move on the spring wave. Many warblers which nest in the Great Slave Lake region in an average temperature of 47°, linger in the Tropics, and reach New Orleans when the temperature is about 65° F. Then they hasten northwards, outstripping the advancing spring, finding in Minnesota a temperature of about 55°, and 52° in Manitoba, and gain another 5° on the season by the time they reach their home.

Thus they continually reach colder weather as they travel north.

The American robin, *Turdus migratorius*, moves more sedately; it takes seventy-eight days for its 3000 mile trip, whilst spring takes some ten days less to cover the distance. But the individual robins may advance more quickly; it is the robin as a species which takes this time to cover the area of distribution. The isotherm of 35° F., corresponding to the beginning of spring migration, advances north at the rate of 3 miles per day from January 15th to February 15th; 10 miles a day is the average for the next month, and 20 for the following month. But along the eastern foothills of the Rockies, isotherms travel faster than in corresponding latitudes farther east; spring rushes to this western land. In mid-April to mid-June—the height of migration—the southern portion of the Mackenzie Valley has about the same temperature as the region of Lake Superior 700 miles farther south. This, coupled with the diagonal course of the birds across the fast-moving region of spring, exerts a powerful influence upon migration; the earliest robins reach southern Iowa on March 1st, and travelling northward at about 13 miles per day, find in central Minnesota a temperature similar to the one they left. Those which breed near Lake Superior increase their speed to a daily average of 25 miles, and arrive at latitude 52°,

when the temperature is still about 34°. The
isotherm, however, has reached central Athabasca,
and the Mackenzie Valley and Alaska robins double
and quadruple their daily average on the north-west
diagonal to keep pace with the spring (19, 20, 21).

Cooke's estimates of the times of advance have
been severely criticised, but certainly some species
forge ahead and others lag behind the vernal wave.
Each species needs separate tracing in its routes and
times and habits, but on the whole the movements
have relation to the changes in seasonal temperature.
In autumn the journey varies according to the time
of starting. Early fall migrants, and indeed the
majority of autumn migrants all the world over,
travel more slowly than in spring ; they are neither
impelled by sex-impulses nor the need to escape from
failing food supplies. A little later the supply does
slacken and with it the temperature cools, and if
the changes are sudden southward migration is
accelerated. Migration, however, is such an advantageous and well-established habit that it usually
begins before hurry is necessary, and the birds loiter
southward, feeding as they go.

Mr Cooke shows that in spring, weather seldom influences the start from the winter home, but the
average weather conditions regulate the *average* rate
of northward advance and the date of arrival at
the breeding home (22).

CHAPTER VIII

THE PERILS OF MIGRATION

THE dangers to which migratory birds are subjected during their journeys are but little less than those which would befall them if they remained in unsuitable zones. During long oversea passages fatigue and hunger weed out the weaklings, sudden storms and adverse winds strike them where no land is near, and they are carried often far from the goal they aimed at. Predatory birds accompany them, taking toll *en route*, and predatory man waits for the tired wanderers with gun and net. Shore birds may rest upon the waves ; sandpipers have been seen feeding as they walked upon the drifting weed of the Sargasso Sea, and steamers and other vessels frequently provide a rest for weary birds ; but what happens to the many which find no haven ? " Woe to the luckless warbler whose feathers once become water-soaked !—a grave in the ocean or a burial in the sand of the beach is the inevitable result," says Mr Cooke. A storm on Lake Michigan during spring migration piled many birds along the shore, and in the wider Gulf of Mexico many hundreds of passage birds were seen to fall into the water when

caught, but 30 miles from land, by a violent
" norther." Other similar sudden disasters have
been recorded off our British coasts, even so far back
as 1786, when, as quoted by Southwell, a Newcastle
collier passed through water off the Suffolk shores
black with vast numbers of drowned woodcocks.

During normal migration birds may be brought to
a lower elevation by strong contrary winds, or they
may be bewildered by fogs and cloud and dropped
nearer the surface ; it is then that the travellers
meet with disaster at our coastwise lights.

Mr Tomison records some of his experiences of
migration at Skerryvore (52). He never saw a bird
at the windows when the moon was shining, and on
clear nights the passing crowds go on without a
pause. But on hazy nights, with an easterly wind
and drizzle, or during fogs, if large numbers of
migrants are passing, hundreds may be seen flying
in all directions, " all seemingly of the opinion
that the only way of escape out of the confusion—
is through the windows of the lantern." On one
September night, when he was standing on the
balcony, he likens the appearance of the birds to a
heavy fall of snow. " Thousands were flitting about ;
hundreds were striking against the dome and
windows ; hundreds were sitting dazed and stupid
on the trimming paths ; and scores falling to the
rocks below, some instantaneously killed, others

seriously injured, falling helplessly into the sea."
On the following night when many fieldfares, redwings, thrushes and other birds were passing, he
says—" Sometimes we use the terms hundreds and
thousands without thinking what these figures mean
but on this occasion when I say thousands were
killed I do not exaggerate in the slightest."

Mr W. Brewster's account of his experiences
at the Point Lepreaux lighthouse (8), shows that
similar disasters occur in Canada and the States,
as indeed they do wherever there are passages of
birds.   On a foggy evening in September 1885
" as soon as the sky became overcast small birds
began to come about the light—with the advent
of the fog they multiplied tenfold in the course of
a few minutes " and many struck.   " About the
top of the tower, a belt of light projected some
thirty yards into the mist by the powerful reflectors ;
and in this belt swarms of birds, circling, floating,
soaring, now advancing, next retreating, but never
quite able, as it seemed, to throw off the spell of the
fatal lantern. . . . Dozens were continually leaving
the throng " of birds which had flown to leeward,
" and skimming towards the lantern.   As they
approached they usually soared upward, and those
which started on a level with the platform usually
passed above the roof. . . . Often for a minute or
more not a bird would strike.   Then, as if seized

by a panic, they would come against the glass so
rapidly . . . that the sound of the blows resembled
the pattering of hail." During his stay no birds
came to the light except during dense cloud or
fog, and they came in greatest numbers when an
hour or two before the fog the sky was clear.

The experiences of Eagle Clarke, Seebohm and
others who have spent migration seasons at lighthouses might be quoted, but these two give a vivid
description of what regularly takes place when
weather conditions are unfavourable. Steady white
lights are the most fatal to migrants, revolving
lights, if white, are struck by some birds, but red
lights seldom attract the passers. Mr Eagle Clarke
thinks that birds are actually decoyed from their
path and arrested in their course by the action
of the lights; he says that a change from white
to red lights at the Galloper Lightship stopped bird
attraction.

On the mainland a new high building or tower,
new telegraph wires or other erections, until their
presence is familiar, take toll of passage birds.

Mr R. M. Barrington has for years collected
information from the Irish lighthouses and light-
vessels; some of his results were added to the
work of the British Association Committee, and
some he published himself (5). He emphasises the
fact that these phenomena depend largely upon

weather, and therefore are not trustworthy indica-
tions of the density at any time or place of migration.
Out of 115 song thrushes killed at the lights and
sent to him, 80 per cent struck during the fourth and
first quarters of the moon, and the same rule holds
good for other species. The intimate relation
between the lunar phases and the number of examples
killed was shown by statistics from 1888 to 1894.
Out of 673 specimens received only 116 were killed
when the moon was more than half full.

Apart from fog or cloud, birds may fail to hit the
land aimed at, either through accidental divergence from correct direction or wind drift.   In
November 1884 Mr Barrington received information
of large numbers of rooks passing simultaneously
at the Tearaght and Skelligs Lights — island
stations 20 miles apart and each 9 miles off the
Kerry coast.   The birds arrived in continuous
flocks from the westward—the open Atlantic—
and passed in an easterly and landward direction ;
they came in small parties and in flocks numbering
two or three hundred, on many days between the
2nd and 25th of the month.   A few birds were
noticed at the same time at stations on the south
and east Irish coasts, and all alike making for the
land.   From similar observations made in other
years he concludes that these were portions of hosts
which had overshot the mark, and failing to find

land had turned back. The weather charts, he adds, show no sufficient reason for the birds to have been blown out of their course by storms.

The weather charts, as I have pointed out, do not indicate the force or direction of the wind at high altitudes ; I suggest that these birds were carried rather than blown out of their way by strong currents at a higher altitude than recorded on the charts, and that having left the air currents they descended to the elevation of about 700 or 800 feet at which most of them were flying when they were observed making for the land.

On the night of March 29th to 30th, 1911, the south-eastern extremity of Ireland experienced a remarkable rush of migrants, and the local papers were full of the avian disaster, for large numbers of birds struck the lights as well as buildings and other objects in inland towns. Mr Barrington collected information (4), and found that most of the birds were starlings, though thrushes, blackbirds, and redwings were numerous. He received specimens of woodcock, water-rail, snipe, dunlin, meadow pipit, wheatear, goldcrest, starling, song thrush, redwing, blackbird, black redstart, robin, skylark, and stonechat, whilst some thirteen or fourteen other species were said to have been recognised, amongst them oyster-catcher and wild duck. The area affected lay south-east of a line drawn across country from Balbriggan to the Old Head of Kinsale, with a coast line of some 200 miles ; most of the birds noticed inland were at towns on the rivers Suir, Barrow, and Nore. The flight was mostly north-east, and at the lights offshore, towards the land. Mr Barrington gives the following explanation. After crossing the Channel the coast of Wexford was reached and the stream divided, some going north along. the east coast and others westward along the south coast, but changing their direction when they reached the wide mouth of the Barrow. The flocks which passed Lucifer Shoals, 10 miles offshore, proceeded north without touching Wexford. Northerly and easterly winds had prevailed for weeks prior to the 29th over France and the British Islands, and birds would be held up in southern Europe ; the milder coastwise temperature of western France, he thinks, would cause them to take a more westerly course than usual. On the morning of the 29th the wind changed to the south at Valentia, Pembroke and the Scilly Islands, and there was an average rise of 7° in temperature at French stations. This rise and the southerly wind liberated the birds, but as the wind continued north-east or east in England they " decided " to take a longer and more exhausting course than usual, pass to Ireland and then turn north-east. The change took place exactly on the last day of the last quarter of the moon—the darkest night for travel. A bank of fog and drizzle met them off the Irish coast, and baffled and weary they were attracted by the lights, not only on the coast but in the inland towns they passed.

In the main I think Mr Barrington's explanation is correct, but even if the birds were gathered farther west than usual, which I doubt, it was the north-east wind which had drifted them, and the word " decided " is a bold one to use when dealing with the behaviour of birds. Easterly winds would drift them westward, and the striking Ireland was accidental ; it was the safety of the many, as well as the deathblow to the comparative few. On the night of the 31st I received news of this visitation, and later found that similar movements, without disaster, were noticed on the north coast of Wales and in Cheshire. On the nights of the 30th and 31st birds in large numbers passed over Bangor and the Menai Straits ; amongst them were golden plover, and the next day these birds with fieldfares and redwings were more abundant than before in the mid-Cheshire fields. On the night of April 2nd, from dusk to midnight, a large passage occurred over Mere in Cheshire, where curlew, golden plover, oyster-catcher and wild duck were recognised by their calls, and at the same time a passage was observed at Old Colwyn on the Welsh coast. I do not even suggest that these were the same birds which passed over south-eastern Ireland, but their presence within so short a time, indicates the volume of the movement.

Welsh papers recorded an "Extraordinary feathered catastrophe" at Pwllheli in Cardigan Bay which occurred on the night of March 17th, 1904, in which "thousands" of birds fell dead and dying upon the town and shore. The journalistic description was lurid, but I am able to give the explanation sent to me by a friend who was an eye-witness. The night had been dark and foggy, and in the morning he found "scores of dead starlings, redwings, thrushes and blackbirds lying on the beach at high-water mark." During the night a steamer had been loading setts at the quarry at the Gimlet Rock, a large outcrop outside the harbour, and the artificial light used had been one of the powerful oil flares. The fog-bewildered birds were led astray and had struck masts, rigging, and rock in their confusion.

During a big fire in Philadelphia on March 27th, 1906, Mr W. Stone saw large numbers of birds passing in its illumination, and many passed too near and fell into the blaze ; he picked up a few half-burnt song sparrows and juncos.

Blizzards on continents, and to a less extent snowstorms in our islands, account for the death of thousands of travellers. And even in most favourable weather birds fall exhausted. During a stay on the Yorkshire coast in autumn, when migration was evenflowing and unchecked by adverse weather, I found several goldcrests which had reached land only to die, and though most birds came in without showing signs of fatigue, a few larks and starlings were so tired that they made little effort to escape when approached.

Ornithological literature supplies many accounts of more or less similar disasters to migrating birds, but these are enough to show that the perils of migration are not exaggerated.

H

CHAPTER IX

EARLY IDEAS OF MIGRATION

THE evolution of the study and knowledge of migration is an interesting subject, dealt with more or less completely by several writers. In a manual it is impossible to treat it fully. That the Greek poets—Homer and Anacreon for instance, and the writers of Jeremiah and Job, knew something about the regular movements of birds is evident, nor is it surprising that in lands like Greece, Egypt and Palestine the passage of birds should be noted and directly connected in the popular mind with the seasonal changes.

In a measure similar observations and conclusions may be traced in the history or traditions of most peoples, but in a northern detached area, such as the British Islands, there is a marked tendency to overlook passage and note only arrival and departure, mostly of summer birds. Early observers noticed the swallow and cuckoo when they had actually come, and missed them when they had gone, but they failed to grasp whence they came or whither they went. Interchange of ideas with inhabitants of other lands was limited,

and few early travellers were philosophers, at any
rate so far as migration was concerned.   In Germany,
however, the Emperor Frederic II. realised in the
thirteenth century many truths concerning migration (27), but in Britain uncertainty or myth held
sway until the end of the eighteenth century.   Herr
Herman, reviewing the variation in thought, says
—" But as in other fields, this period is followed
by a time of decadence, a natural consequence of
departing from immediate experience."

British, and many Continental observers too,
saw when birds had come and in autumn that they
had gone.   Early swallows and martins were always
met with near water, and were watched dropping
to roost in the reed beds, as they always do in
autumn before departure.   Next morning none
was visible.   Certainly then they had vanished
to hibernate in the water.   The discovery of masses
of torpid swallows, dead or dying, by no means an
unknown thing when birds are overtaken by sudden
falls in temperature in autumn or by a severe set-
back in the spring, was to these puzzled men
confirmation of their theory of hibernation.   Other
details of the many stories of swallow hibernation
are due to exaggeration or to misconception.   In
the second half of the eighteenth century a fierce
discussion waged for and against hibernation, and
many, including Geoffroy St Hilaire and Montagu,

sat on the fence, admitting that it might be possible
with some species and probably was with swallows.
Later some Americans produced "evidence" in
favour of avian hibernation, and even Mr Charles
Dixon, in his earlier book at any rate, did not think
it impossible (25). The only argument in favour
of hibernation is that it is a habit resorted to by
other vertebrates to escape the consequence of
exposure to severe temperatures. The arguments
against it are that not a single instance of avian
hibernation will stand the light of reason and investigation, and that birds are provided with
the means of escaping from the cold zone and certainly use these means. There are flightless birds,
but they all live in climates in which they can exist
at all seasons. As Seebohm puts it—" The hibernation of birds is a theory, the evidence in support
of which has completely broken down. The migration of birds is a fact, as completely authenticated
as the fact of their existence."

Dr Derham's "Physico-Theology" appeared in
1737 (24), and contained some sound reasoning
about migration, though he was a little puzzled
with the many hibernation stories. In 1780 an
anonymous pamphlet—"A Discourse on the
Emigration of British Birds," flouted the theory
of winter sleep in no measured terms (33). This
pamphlet was, at first, attributed to George Edwards,

and the 1811 edition has his name on the title, but Mr A. C. Smith shows that the real writer was a comparatively unknown man, John Legg. Legg must be looked upon as one of our first real students of migration. It is Legg who refers to a pamphlet which appeared in 1740 in which it was seriously argued that swallows migrated annually to the moon.

All this time, from 1736 onwards, the family of Marsham in Norfolk, had been quietly recording observations on the arrival of migrants, each generation continuing the work. The accumulated results have been used, and will be used again, in studying the science of " ornithophaenology."

A myth, founded on mistaken observation as well as upon mere speculation was, and to some extent still is, that the larger migrants assist the passage of the weaker ones. How else, is still asked, can weak-winged species cross the sea ? It was an old legend when J. G. Gmelin heard it from the Tartars in 1740 ; each crane they told him took a corncrake on its back. There are men who know the corncrake well, who believe to-day that the bird must skulk unseen through the winter, for they assert it is quite incapable of lengthy flight. It is useless to argue with them ; the only answer is that it not only can, but regularly does perform a long double journey ; its range extending from northern Europe to South Africa. In 1911 I handled a water-rail, a bird with short rounded wings like those of the corncrake, which had struck the lantern of a lighthouse with great violence. Its smashed head was nearly severed from its body.

Herr Otto Herman's " Recensio critica automatica " (31) supplies much information about the literature on bird migration, and the strange divergence of opinion on nearly every point. It is carried up to the beginning of the twentieth century, but much of the valuable work done in America is altogether neglected.

A short bibliography is given at the end of the present volume, including the more important works on the subject and a few of the papers in periodical publications referred to in this manual.

CHAPTER X

SUGGESTIONS AND GUESSES

SEVERAL important migration phenomena have hardly been touched upon in the previous pages ; a few words about these may not be out of place.

There is no doubt that now and again American species are met with in Europe, and European in America, though there is no evidence of direct regular trans-Atlantic passage, except from Greenland. The appearance of these birds has been explained in several ways, the general notion being that it is impossible for a bird to fly unaided across the Atlantic, say over 3000 statute miles, without rest. In considering the question we are met with various points on which we still lack knowledge.

We know that strong-winged waders can accomplish 2500 miles, apparently without a rest, and that if rest is necessary these birds can swim and rise from the waves. We know, too, that there is regular passage between Greenland and Europe. We do not know how long a bird can, without rest and food, sustain flight ; we do not know the speed it can travel when aided by favourable winds, nor to what extent even passerine birds may rest upon the water. My friend Mr J. A. Dockray, when punting in the Dee estuary, has often seen birds alight to rest on his punt, and once saw a tired thrush settle repeatedly on the water and finally safely cross the estuary. There are several instances recorded of passerine birds alighting upon and rising again from the water.

We do not know the extent of Greenland as a summer breeding home of birds; the growing knowledge of this vast continent proves that its summer avifauna is much larger than we thought, and that western and eastern forms inhabit adjacent breeding areas; the possibility of birds banding with the wrong set of travellers is greater than was suspected.

It is urged that the western shores of Scotland and Ireland should receive these stragglers, but that the records of American birds are fewer from these coasts than from the eastern shores and even Heligoland. The best island route, however, would lead birds to join the travellers from Scandinavia which pass by the safer eastern route than the one round the western wind-swept shores of Ireland. Even this reputed scarcity may be error, for how many reliable watchers are there compared with the immense length of this wave-indented coast-line ? How easy for a straggler to be overlooked ! Mr S. F. Baird, in his paper on the " Distribution and Migration of North American Birds," is emphatic that the transfer of American birds to Europe is entirely due to the agency of winds carrying them from their course (6). Mr A. L. Butler met with snow - buntings in mid - Atlantic travelling east, and Mr J. Trumbull supplies information about many passerine birds — especially snow - buntings and wheatears — seen in September and October at various points between Canada and the British coasts (53). Some joined ships but others made no attempt to do so, even at 54° north 44° west.

Unfortunately there is the negative evidence of fraud, for when unscrupulous dealers found that the public would give high prices for rare birds, a trade in American skins began. It is not impossible that even Gätke was victimised. Error or even accidental fraud may be taken into account. Some years ago I heard that a hawk-owl had been killed in Cheshire, at an inland port on the Ship Canal ; I traced the bird, the American species, but discovered that it had been captured on an east-bound steamer in the Straits of Belle Isle, and had only died or been killed when the vessel reached the coaling station at Partington, where the taxidermist who received it thought it had been taken. A Cape pigeon, which I saw in the flesh, reported as shot in Lancashire, I found had been brought home in cold storage.

Birds may be carried on ship-board. When the
" Mauretania " was between 400 and 500 miles
out from New York, bound eastward on June 15th,
1911, a curlew came on board and remained for
three days, leaving when the Irish shores were
sighted on the 18th. My informant, an experienced wildfowler, failed to catch the bird,
but described it as like our curlew. Probably
it was the American *Numenius longirostris*, but
amongst the Irish curlews it would easily remain
unrecognised.

When a seabird appears inland the usual explanation given is " storm-blown," but increasing
knowledge shows the frequent fallacy of this idea.
The Manx shearwater, for instance, is a regular
migrant, and the examination of the dates of the
records of so-called " storm-blown " birds found
in inland localities, shows a remarkable regularity ;
the majority are met with between the end of
August and the end of the first week in September.
Not only do the birds move south in the early days
of September but many, usually at any rate, cross
England ; the weaklings fall out and are found.
Is it possible that some of these collapses of
passing birds are due to more than mere physical
fatigue ? Aviators have discovered the existence
of " wind pockets " or " holes in the air," where
the resistance of the air appears suddenly to fail ;

what is the effect on a flying bird which suddenly enters one of these pockets ?

The lesser black-backed gull also crosses England in large numbers ; its movements are more noticeable than those of the herring gull, common gull, or even of the inland nesting and inland feeding black-headed gull.

Recent investigation has added the yellow-browed warbler, the blue-throat, and many other " rare," or " casual " passerine birds to the list of regular British birds of passage ; evidently they have been overlooked before. Even the crossbill, so long classed as a spasmodic invader, is now seen to be a regular bird of passage to Britain, though in varying numbers, and quite independently of the sub-specific form which is always with us.

The wanderings to our islands of southern petrels and other oceanic birds has occasioned much surprise. Take two examples of the genus *Oestrelata*, one *O. brevipes* taken at Borth in 1889, and *O. neglecta* in Cheshire in 1908, the known breeding range of both being in the western Pacific ; pelagic wanderings might lead a bird anywhere, but it is conceivable that investigation may show that the breeding area is wider than is supposed and that these species have stations even in the South Atlantic.

Some writers affirm that birds only migrate on the wing, but the journey by sea of many species is varied in method. Those very regular migrants, the puffins and guillemots, which the light keepers assure us leave and return to their stations almost at fixed dates, move by slow nautical stages, swimming and feeding as they go. On May 2nd, 1911, I watched a red-throated diver slowly travelling north ; it actually travelled farther beneath the surface than either by swimming or flying, so long as I had it in view. The penguin's migrations cannot possibly be on the wing. Dr Brooks rightly contended that the periodic assemblage of wandering sea-birds at their " rookeries " is true migration, regular as the almanack, although the feeding area is immense and the birds do not reach home by any single path. Seebohm tells us of young bean geese migrating in full moult, marching in an army to the interior of the Tundra, and Mr W. H. Hudson, in " Birds and Man," relates a pathetic story of a pair of upland geese in southern Buenos Ayres. His brother saw them in August, the early spring of South America, leaving the plains where they had wintered to breed in Magellanic islands. The main flocks had departed, but these two birds, the female with a broken wing, were steadily moving south, the male taking short flights and waiting for her, as if to urge her on, and the female walking. "And in this sad, anxious way they would journey on to the inevitable end, when a pair or family of carrion eagles would spy them—and the first would be left to continue the journey alone."

That strong-winged birds can cross the Atlantic is proved by the appearance in December 1927 of many lapwings in Newfoundland and Labrador, where the bird was unknown except as a very rare straggler. One bird, shot at Bonavista on December 27th, carried a ring placed on its leg at Ullswater, Cumberland, when a nestling in May 1926, by my friend, Dr H. J. Moon. Strong westerly winds and severe cold towards the end of the month occasioned large south-westerly weather migrations of ground-feeding birds. On the morning of December 20th the wind, due east to west, reached at 1000 feet an average force of 55 miles per hour to within 200 miles off the Newfoundland coast. This Mr Witherby learnt from the Air Ministry, who said it would still be strong at much lower altitudes. Adding this speed to the normal pace of the lapwing, 45 miles per hour, we see that the 2200 miles could be covered in under twenty-four hours, even allowing for reduction of speed as the coast was approached. At any rate, large numbers of lapwings landed in all parts of Newfoundland, the flocks recorded on varying dates and in different places numbering from about a score to several hundred birds.

CHAPTER XI

SUMMARY

MIGRATION owes its origin to the potentiality of flight, enabling birds to advantage themselves by extended dispersals, which through heredity become instinctive, regular and periodical. Geological changes, especially the passing away of the glacial epoch, only influenced by opening up new lands for summer colonisation, but climatic conditions prevented these lands from becoming permanent abodes and fostered the habit of periodical migration. Whatever the original home or centre of distribution may have been, the dispersal from it was towards new lands with a retreat towards the food-supply when these lands became untenable. Fluctuating food-supply, love of home, sexual impulses, desire for light, varying temperature, and other factors, all have more or less influence, but the force exerted by any or all depends upon the species operated upon and the locality in which it resides. The present route followed or method of migration is little guide to the history of past migration ; during the evolution of present-day migration alterations may have been occasioned by
126

environment and changing conditions. As Seebohm puts it, " The desire to migrate is a hereditary impulse, to which the descendants of migratory birds are subject—a force almost, if not quite, as irresistible as the hereditary impulse to breed in the spring " (44).

The route is simply the course followed between the breeding area and winter quarters ; it is more or less restricted by the size of the area in which food is to be found ; it is usually the most direct way from one food-base to the next, in a general direction from the seasonal bases. Most birds move between north and south, but migrations are regularly followed in other directions by some species.

Routes may follow coastlines, these providing visible landmarks, and also, for many species, plentiful food ; islands, capes, estuaries and inlets are landmarks, asylums, food-bases, and sites for congregation and departure for cross-sea passages ; at these places migration is often specially noticeable. Overland routes may suggest " broad front " migration, when there are no particular restricting influences and the species have no special need for hurry. Migration at great elevations and at high rates of speed is proved, but the highest and quickest possible is as yet unascertained. It may also, under other conditions, be performed at low elevations and very slowly. It is probable that strong air-currents at a high elevation materially assist rapid and lengthened migration. Force not direction of wind influences birds moving at a low elevation.

Birds possess a certain power of orientation, a homing instinct, which need not be called a sixth sense. Brain and eyes assist in the development of this power; birds have an excellent memory. Young birds lose their way more frequently than is generally supposed; variations in routes are explained in many cases by these errors. Young may or may not be guided by experienced adults; orientation is not infallible but develops with age.

There is apparently no truth in the assertion that birds travel by choice against a head wind or in a beam wind; a moderate wind behind, on which they are carried, is most favourable. Leeward drift through contrary winds explains many normal and abnormal routes, and the occurrence of un-expected species in unexpected places. The distance travelled not only varies according to species but in individuals of the same species; the thesis that the most northerly breeder winters farthest south does not always hold good.

Much may be learnt by the careful registration of arrivals and departures of migratory birds, and by the marking of birds. Ornithophænology, the science of migration study, as carried on at present

in many countries, would be materially assisted
by some better method of international registration and interchange of ideas.

In conclusion I would urge the value of the study,
citing Herr Herman's reasons put before the International Ornithological Congress in 1905. The
solution of the problem is in the interest of science,
and therefore of intellectual progress, teaching
us the great part which migratory birds play in
the scheme of nature. The millions of birds which
wander, season after season, from one zone to
another, represent an enormous aggregate of labour,
by flight and search for food, acting on " the organic
life of nature as does the regulator of a steam-engine,
at one time accelerating, at another retarding."
Full insight into the essence of the work done by
birds will give us a correct notion of their usefulness
or injuriousness to man, and lead us to rational
action for their protection.

Whilst fully agreeing with Herr Herman I would
go further. We live in an age when aerial locomotion has become important, and will be more and
more important in the future. Every lesson we
can learn from the successes or failures of these
most perfect aerial navigators must be of use.

But putting aside economic and utilitarian con-
siderations, there is to some of us a greater stimulus
to solve the problems of nature. With the birds,

I

and the insects and plants upon which they feed, we share a common heritage, and the more we learn of the life of these, our fellow workers, the nearer we approach solution of the great riddle of the Universe, the mysterious law-abiding scheme of Nature. The book of knowledge to which we may add some iota is marred with mystery, superstition and error, but each proved fact cleanses its pages. " Facts," says Laing, " are the spokes of the ladder by which we climb from earth to heaven."

BIBLIOGRAPHY

1. ALLEN, J. A.   Cooke's *Some New Facts about the Migration of Birds, Auk*, xxi., 1904, 501.
2. —— Gätke's *Heligoland, Auk*, xiii., 1896, 137.
3. —— Walter's *Theories of Bird Migration, Auk*, xxv., 1908, 329.
4. BARRINGTON, R. M.   "The great rush of Birds, etc." *Irish Nat.*, xx., 1911, 97.
5. —— *The Migration of Birds as observed at Irish Lighthouses and Lightships*, London, 1900.
6. BAIRD, S. F.   "The Distribution and Migration of North American Birds." *Amer. Jnl. Science and Arts*, 2, 1866, xli.
7. BREHM, C. L.   "Der Zug der Vögel," *Isis*, 1828, *Naumannia*, 1855.
8. BREWSTER, W.   "Bird Migration." *Mem. Nuttall Orn. Club.* Cambridge, Mass. No. 1, 1886.
9. BROOKS, W. K.   *The Foundations of Zoology*, New York, 1899.
10. *Bulletin of the British Ornithologists' Club. Reports on Migration*, vols. xvii., xx., xxii., xxiv., xxvi., 1906-1910.
11. CARPENTER, F. W.   "An Astronomical Determination of the Heights of Birds," *Auk*, xxiii., 1906, 210.
12. CHAPMAN, ABEL.   *Bird-Life of the Borders*, 2nd edit., London 1907.
13. —— F. M.   "Observations on the Nocturnal Migration of Birds," *Auk*, 1888, 37.
14. CLARKE, A. H.   "The Migration of Certain Shore Birds," *Auk*, xxii., 1905, 134.
15. CLARKE, W. E.   "Bird Migration in Great Britain." *Report of the British Association, London*, 1896.
16. —— "Studies in Bird Migration," *Ibis*, 1902, 246, 1903, 112.

I*   131

17. CLARKE, W. E. "The Birds of Fair Island; Native and
       Migratory." *Ann. Scot. Nat. Hist.*, 1906, 4.
18. COOKE, W. W. "Distribution and Migration of North
       American Shorebirds." *U. S. Dept. Agric.
       Biol. Survey*, Bull, 35, Washington, 1910.
19. ——— "Routes of Bird Migration," *Auk*, xxii., 1905, 1.
20. ——— "Some New Facts about the Migration of Birds." *U. S.
       Dept. Agric. Year Book*, 1903, 371.
21. ——— "Our Greatest Travellers." *Nat. Geog. Mag.*, 1911, 346.
22. ——— "The Migratory Movements of Birds in Relation to the
       Weather." *U. S. Dept. Agric. Year Book*,
       1910, 379.
23. CORDEAUX, J. "Migration in the Humber District," *Zool.*,
       1892, 418.
24. DERHAM, W. *Physico-Theology*, London, 1737. Lect. de-
       livered in 1711-12.
25. DIXON, C. *The Migration of Birds*, London, 1892.
26. ——— *The Migration of British Birds*, London, 1895.
27. FREDERICK II., (Emperor). *De Arte Venandi cum Avibus*,
       Ed. Schneider, 1788, (Rhea. ii., 1849).
28. GADOW, H. F. "Migration," *Encyclo. Brit.*, 11th Edit.,
       Cambridge, 1911.
29. GÄTKE, H. *Heligoland as an Ornithological Observatory*,
       Trns. Rosenstock, London, 1895.
30. HERMAN, O. "A.M.O.K. Ornithophænologiæ anyaja,"
       Aquila, 13, 1906, xx.
31. ——— *Recensio Critica automatica of the Doctrine of Bird-Migra-
       tion*, Budapest, 1905.
32. LAIDLAW, T. G. "Reports on the Movements and Occur-
       rences of Birds in Scotland during 1902 and
       1903." *Ann. Scot. Nat. Hist.*, 1903-4.
33. (LEGG, JOHN). *A Discourse on the Emigration of British
       Birds*, London, 1795. (Salisbury, 1780, and
       London 1811, the latter under name of
       George Edwards.)
34. LINNÉ, C. *Dissertatio migratione Avium*. Upsaliae, 1757.

35. MIDDENDORF, A. T. VON. *Die Isepiptesen Russlands Grund-lagen zur Erforschung der Zugzeiten und Zugrichtungen der Vögel Russlands*, St Petersburg, 1853.

36. MENZBIER, M. " Die Zugstrassen der Vögel im Europäischen Russland." *Bull de la Soc. Imp. d. Nat.*, Moscou, 1886, 291.

37. NEWTON, A. *A Dictionary of Birds*, London, 1893-1896.

38. —— " Migration," *Encyclo. Brit.*, 9th Edit., London.

39. PALMÉN, I. A. *Om foglarnes flyttingsvägar*, Helsingfors, 1874.

40. —— *Über die Zugstrassen der Vögel*, Leipzig, 1876.

41. SCHÄFER, E. A. " On the Incidence of Daylight as a deter-mining factor in Bird Migration." *Nature*, 1907, 159.

42. SCLATER, W. L. " The Migration of Birds in South Africa." *S. African Orn. Union*, 1906, II., 14.

43. SCOTT, W. E. D. " Some Observations on the Migration of Birds." *Bull. Nuttall Ornith. Club*, vi. 97.

44. SEEBOHM, H. *Geographical Distribution of the Family " Chara-driidae*," London, 1888.

45. —— *The Birds of Siberia*, London, 1901.

46. SERVICE, R. " Bird Migration in Solway." *Ann. Scot. Nat. Hist.*, 1903, 193.

47. STEBBINS, J. and FATH, E. A. " The use of Astronomical Telescopes in determining the speeds of Migratory birds." *Science* (New York), xxiv., 1906, 49.

48. STEJNEGER, L. " Do Birds Migrate along their Ancient Im-migration Routes." *Condor*, vii., 1905, 36.

49. STONE, W. " Bird Waves and their Graphic Representation," *Auk*, 1891, 194,

50. STUBBS, F. J. " The Use of Wind by Migrating Birds." *Mem. and Proc. Manchester Lit. and Phil. Soc.*, vol. 53, 1909.

51. TAVERNER, P. A. " A Discussion of the Origin of Migration," *Auk*, xxi., 1904, 322.

52. TOMISON, J.   " Bird Life as observed at Skerryvore Lighthouse."   *Ann. Scot. Nat. Hist.*, 1907, 20.
53. TRUMBULL, J.   " Notes on Land Birds observed in the North
Atlantic and the Gulf of St Lawrence,"
1904.   *Zoologist*, 1905, 293.
54. WALLACE, A. R.   *Nature*, x., 1874, 459.
55. WHITLOCK, F. B.   *The Migration of Birds*, London, 1897.
In addition numerous notes in the following periodicals have
been consulted :—*Annals of Scottish Natural
History, Auk, British Birds, Condor, Emu,
Field, Ibis, Irish Naturalist, Naturalist,
Nature, Zoologist.*
56. VERY, F. W.   *Science*, n.s. VI., 1897.
57. WATSON, J. B., and LASHLEY, K. S.   " Homing and Related
Activities of Birds."   *Dept. of Marine Biology*,
Washington, 1915.
58. THOMSON, A. L.   *Problems of Bird-Migration*, London, 1926.

The above list is of books and papers specially referred to, but
further information may be obtained about Dr Watson's experiments
in *Bird Lore*, 1908, 134 ; on migration in Scotland in the annual
reports of Mr T. G. Laidlaw and the Misses E. V. Baxter and L. J.
Rintoul, published in the *Annals of Scottish Natural History* and
the *Scottish Naturalist*; and on the movements of certain species in
the British Association Reports from 1900 to 1903.   The substance
of these last-mentioned reports, largely the work of Mr Eagle Clarke,
his papers referred to above (15, 16, 17,) and many others, have been
revised and incorporated in his " Studies in Bird Migration " (London,
1912, 2 vols.).   A very important work on the subject is Dr A. L.
Thomson's *Problems of Bird-Migration.*

INDEX

www.ingramcontent.com/pod-product-compliance
Ingram Content Group UK Ltd.
Pitfield, Milton Keynes, MK11 3LW, UK
UKHW010851090126
466816UK00011B/152